Tourism Management in Warm-water Island Destinations

CABI Series in Tourism Management Research

General Editors:
Professor Eric Laws, Professor of Tourism, Siam University, Bangkok
Professor Noel Scott, Griffith Institute for Tourism, Griffith University, Australia

Since the mid-20th century, modern tourism has grown rapidly in extent and diversity, becoming increasingly competitive and volatile as it is impacted by climate change, new technologies, changing distribution systems and the opening of new markets. As a result, governments, tourism destinations and businesses need to improve their management capability and adopt best practices to survive. The purpose of this series is to provide tourism managers, administrators, specialists and advanced students with state-of-the-art research and strategic knowledge to enable them to thrive in dynamic and unpredictable environments. Contributions are based on critical and interdisciplinary research that combines relevant theory and practice, while placing case studies from specific destinations into an international context. The series presents research on the development and diffusion of best practice in business and destination management that fulfils the objective of environmental, sociocultural and economic sustainability at both the local and global scale.

The cover design for this series depicts a generalized mosaic composed of many tiles. Metaphorically, this illustrates our philosophies that while the various elements of tourism require specific study, it is the overall picture that is most significant, and that tourism is a very dynamic, complex and evolving industry. This series seeks to build a coherent approach to future tourism research through each individual title.

Titles available

1 *Tourism Crisis and Disaster Management in the Asia-Pacific*
Edited by Brent W. Ritchie and Kom Campiranon

2 *Destination Competitiveness, the Environment and Sustainability: Challenges and Cases*
Edited by Andres Artal-Tur and Metin Kozak

3 *Visitor Management in Tourism Destinations*
Edited by Julia N. Albrecht

4 *Sustainable Island Tourism: Competitiveness and Quality of Life*
Edited by Patrizia Modica and Muzaffer Uysal

5 *Visitor Experience Design*
Edited by Noel Scott, Jun Gao and Jianyu Ma

6 *Tourism Management in Warm-water Island Destinations*
Edited by Michelle McLeod and Robertico Croes

Tourism Management in Warm-water Island Destinations

Edited by

Michelle McLeod

The University of the West Indies, Mona, Jamaica

and

Robertico Croes

University of Central Florida, Florida, United States of America

CABI

CABI is a trading name of CAB International

CABI
Nosworthy Way
Wallingford
Oxfordshire OX10 8DE
UK

Tel: +44 (0)1491 832111
Fax: +44 (0)1491 833508
E-mail: info@cabi.org
Website: www.cabi.org

CABI
745 Atlantic Avenue
8th Floor
Boston, MA 02111
USA

Tel: +1 (617)682-9015
E-mail: cabi-nao@cabi.org

A catalogue record for this book is available from the British Library, London, UK.

Library of Congress Cataloging-in-Publication Data

Names: McLeod, Michelle T., editor. | Croes, Robertico R., editor.
Title: Tourism management in warm-water island destinations / Edited by Michelle McLeod and Robertico Croes.
Description: Boston, MA : CAB International, [2018] | Series: CABI series in tourism management research | Includes bibliographical references and index.
Identifiers: LCCN 2017023569 (print) | LCCN 2017055322 (ebook) | ISBN 9781786390936 (ePDF) | ISBN 9781786390950 (ePub) | ISBN 9781786390929 (hbk : alk. paper) | ISBN 9781786390936 (e-book) | ISBN 9781786390950 (e-pub)
Subjects: LCSH: Tourism--Management. | Islands--Economic conditions.
Classification: LCC G155.A1 (ebook) | LCC G155.A1 T5923735 2018 (print) | DDC 910.68--dc23
LC record available at https://lccn.loc.gov/2017023569

ISBN-13: 978 1 78639 092 9 (hbk)
978 1 78639 093 6 (e-book)
978 1 78639 095 0 (e-pub)

Commissioning editor: Claire Parfitt
Editorial assistant: Emma McCann
Production editor: Marta Patiño

Typeset by SPi, Pondicherry, India.
Printed and bound in the UK by CPI Group (UK) Ltd, Croydon, CR0 4YY, UK.

Contents

Editors and Contributors

Editors

Robertico Croes is a full professor who currently serves as the Associate Dean of Administration and Finance in the Rosen College of Hospitality Management at the University of Central Florida. He is also the Associate Director of the Dick Pope Sr. Institute for Tourism Studies. His fields of interest include: econometric applications in tourism, demand forecasting, economic impact, sustainable development, poverty alleviation and tourism development for developing countries. He has authored three books and has more than 100 published works. His research has been presented and applied in countries such as Armenia, Malta, Netherlands, Spain, Italy, Aruba, Curacao, Barbados, the Bahamas, Bonaire, Puerto Rico, Venezuela, Nicaragua, Costa Rica, Colombia, Brazil, Ecuador, England, Mexico, South Africa, South Korea, Hong Kong, Ireland, Malaysia, Turkey, Taiwan, Cyprus and Trinidad and Tobago. He is the recipient of numerous research awards such as the 2015 Thea Sinclair Award. E-mail: robertico.croes@ucf.edu

Michelle McLeod is Acting Programme Director at the Centre for Hotel and Tourism Management, University of the West Indies, Nassau, Bahamas. Her tourism industry experience spans over 25 years and she is a graduate of the University of the West Indies, the University of Surrey, and Bournemouth University where she obtained a doctorate with no corrections after her Viva. Her present research interests are in the subject areas of tourism, knowledge networks, open data ecosystems, policy networks, destination networks and service productivity. E-mail: michelle.mcleod@uwimona.edu.jm

Contributors

Sheree-Ann Adams is a PhD graduate from the International Centre for Research in Events, Tourism and Hospitality at Leeds Beckett University in the UK. She works as a consultant, researcher and trainer specialising in responsible and sustainable tourism development, management and marketing as well as a small business shore-excursion operator in the Caribbean region. E-mail: grenscot@hotmail.com

David Airey is Emeritus Professor at the University of Surrey in the UK. His research interests focus on tourism policy and education. E-mail: d.airey1@btinternet.com

Donna Chambers is Professor of Tourism, and Research Lead in the Department for Tourism, Hospitality and Events, University of Sunderland, UK. She is interested in how we represent peoples and places through cultural and heritage tourism, the link between heritage and national identity, postcolonial and decolonial epistemologies, visuality in tourism, and also in critical and innovative approaches to tourism research and scholarship. She is a resource editor for *Annals of Tourism Research* and is on the editorial board of the journal *Leisure Studies.* She also serves as a regular reviewer for a range of other tourism journals. She is currently conducting research on decolonisation and tourism. E-mail: Donna.Chambers@sunderland.ac.uk

Anne P. Crick is a senior lecturer in the Mona School of Business and Management at the University of the West Indies (UWI). She is a graduate of the UWI, Penn State University and Rutgers University. Her research interests include tourism, hospitality management, service management and emotional labour and she has published over 20 research articles and book chapters in those research areas. E-mail: anne.crick@uwimona.edu.jm

Frederic Dimanche has worked on consumer behavior and tourism marketing and development in the USA and Europe. He recently published a book on tourism management and development in Russia. E-mail: fdimanche@ryerson.ca

Rachel Dodds' work focuses on practical, applied management and marketing of tourism. Her main research focus encompasses sustainability issues in tourism. She has visited Cuba a number of times in the past three decades. E-mail: r2dodds@ryerson.ca

Brigitte Joubert is a lecturer and senior coordinator at the Seychelles Tourism Academy (STA). She has a Masters in Sustainable Tourism Planning from Bournemouth University, UK. She is currently based in the Hospitality Management Department at STA. Her research interests are small island developing states and sustainable tourism. E-mail: brigej@gmail.com

Jorge Ridderstaat is currently an assistant professor at the Rosen College of Hospitality Management at the University of Central Florida in Orlando, Florida. Previously, he worked for some 20 years at the Central Bank of Aruba, first as an economist, and in the last 8 years as the Manager of the Research Department. Presently, his research interests are in the area of applied econometrics in tourism and hospitality, tourism demand analysis and forecasting, financial interactions/feasibilities in the hospitality industry and big data analytics. He is the author of two books and has published his research in many top-tier tourism journals. E-mail: Jorge.Ridderstaat@ucf.edu

Manuel Rivera is a professor at the Rosen College in the University of Central Florida. He is the Managing Editor for the International Journal of Hospitality Management, the most regarded hospitality journal worldwide. His industry experience expands to brands such as Aramark, Chartwells, Pizza Hut, Starbucks, Java City, Pollo-Tropical, Subway, Burger King, McDonalds and Sbarro. He has also worked with Florida International University, Saint Thomas University, Nova Southeastern University, Florida Memorial College, the Miami Dolphins Training Center and the Kovens Convention Center. He has carried out research in Puerto Rico, Bonaire, Ecuador, Nicaragua, Curacao, Aruba, Costa Rica, Nicaragua, and Coca-Cola's Latin Center Business Unit. He has vast experience working with the Curacao North Sea Jazz Festival, Aruba Soul Beach Music Festival, Aruba Summer Festival, Curacao Carnival and Aruba Electric Festival. He coauthored the book *Poverty Alleviation through Tourism Development: A Comprehensive and Integrated Approach* and his research has been published in numerous hospitality and tourism journals. E-mail: manuel.rivera@ucf.edu

Michael Sadowski has been actively involved in developing and promoting Intrepid's sustainable tour product. He is part of Intrepid's senior leadership team in North America and a member of the Forbes Communications Council. He is also a graduate of Ryerson University. E-mail: michael.sadowski@intrepidtravel.com

Noel Scott is Professor and Deputy Director, Griffith Institute for Tourism at Griffith University, Gold Coast, Australia. His research interests include the study of tourism experiences, destination management and marketing and stakeholder organization. He is a frequent speaker at international academic and industry conferences. He has over 210 academic articles published, including in 11 books. He has supervised 18 doctoral students to successful completion of their theses. He is on the editorial board of 10 journals and is a member of the International Association of China Tourism Scholars. Prior to starting his academic career in 2001, Noel worked as a senior manager in a variety of leading businesses, including as Manager Research and Strategic Services at Tourism and Events Queensland. E-mail: noel.scott@griffith.edu.au

Neelu Seetaram, PhD, is the head of research and professional practice at the department of tourism and hospitality, Bournemouth University. She is an elected council member of the International Association for Tourism Economics. She is an economist who regularly publishes in the field of tourism economics and has co-edited two books. E-mail: nseetaram@bournemouth.ac.uk

Kelly J. Semrad currently serves as a faculty member in the Rosen College of Hospitality Management at the University of Central Florida. She is the former associate director for the University of Florida's Eric Friedheim Tourism Institute. She possesses 15 years of professional experience in community event planning which includes having served on mega event host committees, such as the World Series and Super Bowl Games. Dr. Semrad has presented over 65 research presentations at international conferences, has published in renowned academic journals, and has served on research teams that have provided evaluations and technical reports to tourism and hospitality organizations worldwide. She is an active team member for numerous grants representing socio-economic needs in developing countries. She is the recipient of numerous research and teaching awards and has also served as an invited visiting professor at multiple universities abroad. Her teaching portfolio is extensive having taught more than 30 different hospitality, sport management and event management courses. E-mail: kelly.semrad@ucf.edu

Vincent Vanderpool-Wallace's experience as co-president of Resorts International, then the largest private company in the Bahamas, as Director General of Bahamas Tourism, as Secretary General and CEO of the Caribbean Tourism Organization and, most recently, as Minister of Tourism of the Bahamas, has made him an ideal advisor to both private companies and governments in the tourism and hospitality arena. He is a strong advocate of policies to ensure robust public-private sector cooperation in the growth and development of businesses and destinations that rely on hospitality for their profitability and their economies. At the end of his involvement in Bahamas public sector tourism, his tenure covered its best performing years whether measured by total visitors, by hotel visitors, by cruise visitors, by tourism revenues or by levels of visitor satisfaction. He has served also as a member of the board of directors for the Central Bank of the Bahamas and as Chairman of the Hotel Corporation of The Bahamas, which was tasked with selling ten government owned hotels and which resulted in the successful establishment of several international brands in the Bahamas. He is a graduate of Harvard University, BA, and the University of Miami, MBA. E-mail: vVw@bedfordbaker.com

List of Figures

List of Tables

1 Overview of Tourism in Warm-water Island Destinations

Michelle McLeod[1]* and Robertico Croes[2]

[1]*The University of the West Indies, Mona, Jamaica;* [2]*University of Central Florida, Florida, United States of America*

1.1 Background and Rationale

Island tourism is a unique form of tourism that requires exemplification. An island is a piece of land surrounded by a body of water. The physical dimensions are varied and these landforms may occur as archipelagos with several islands, atolls, islets and general masses occurring within close range and islands being included based on their appearance during high and low tides. Island formations originate from tectonic activities involving the Earth's crust and these activities provide islands with certain geographical features, such as the Pitons or volcanic plugs in St. Lucia. Island tourism denotes tourism activities within these island environments.

As tourism continues to be a dominant global activity for economic gain and employment, there is a need to understand how the business of tourism affects island environments. Principally, islands are physically resource-constrained with small physical spaces and population sizes. If one was to categorize islands, this would mainly be based on size, with small islands being identified not just by the physical landscape, but more so by having a population of less than 1.5 million (Croes, 2016); and larger islands being those with a greater number of inhabitants. From a resource-constrained perspective, one wonders how the fortunes of tourism growth and development on islands can be managed and maintained.

Tourism activities occur on islands as the resources within islands are natural attractors to tourist markets based on climatic, geographic and cultural features. Islands are some of the most diverse landscapes located in tropical and temperate climates. For ease of identification, although not a strict boundary, warm-water islands are located between the Tropic of Cancer and the Tropic of Capricorn at 23.5° north and south of the equator respectively (Fig. 1.1). Islands that fall within the tropical zone include those in the Caribbean, Indian Ocean and Pacific.

*E-mail: michelle.mcleod@uwimona.edu.jm

Fig. 1.1. World map showing islands between Tropics of Cancer and Capricorn (immediate lines above and below the equator). (From https://www.cia.gov/library/publications/the-world-factbook/graphics/ref_maps/physical/pdf/world.pdf)

The Caribbean is a main hub for island environments and island tourism. The geographical characteristics of islands in the Caribbean can be categorized based on the nature of the rocks, either limestone or volcanic (Boniface *et al.*, 2012). The Caribbean is comprised of islands in the Lesser Antilles as the smaller islands in the Eastern Caribbean, such as Antigua and Barbuda and St. Lucia; and the islands in the Greater Antilles, the larger islands, such as Cuba and Jamaica. The Caribbean region is positioned east of Central America, and stretched out between North and South America. A Caribbean island is so assigned based on its shoreline touching the Caribbean Sea. The region's location to main tourist generating markets is a major advantage for the emergence of tourism activities in the Caribbean. As such, the Caribbean region has a long history of tourism development and management (see Holder, 2013; McLeod, 2015) and is, therefore, an excellent example for considering tourism management principles and practices in island destinations.

Besides the Caribbean, warm-water island destinations are also located in the Indian and Pacific Oceans. The Indian Ocean has some of the most remote islands in the world such as the Maldives, Mauritius and the Seychelles. Tourism has become a main source of income and economic activity; however, the economic dependence on tourism of the islands in the Indian Ocean was put under the spotlight following the 2004 tsunami. The events of the tsunami showed the real impact of physical forces on tourism and the resilience of the tourism sector in the aftermath. The islands of the Pacific are bunched in the South Pacific with a number of islands above the Tropic of Capricorn. According to Apostolopoulos and Gayle (2002, p.6) many islands are microstates that are constrained by resource scarcity and few viable alternatives, hence tourism dependence.

The cultural diversity is also an attractor to island environments. Islands offer inhabitants relaxed lifestyles in culturally enriched areas such as the Caribbean and the Pacific where there is an intermixing of various cultures. Based on African, Asian and European influences, many island destinations have become cultural melting pots with an emergence of distinct cultures that are matched with the national identities of sovereign island states.

Islands are natural tourism hubs as their scenic beauty allures the fantasy-driven motives of tourists, who yearn for a rebirth of existence from mundane and banal lives, through tourist experiences. The shoreline is a main attractor of the island tourism experience. Islands that are smaller in size with close distances to the sea are more favourable to create the common imagery that island tourism provides. Larger distances from the sea have resulted in tourism facilities and amenities being zoned within close proximity to the sea to immerse the tourist and create the fantasy of tourism experiences.

Within the constraints of an island, the dilemma of meeting tourist expectations against the needs of a host population requires balance, and therein is the importance of a tourism management approach to island tourism. The next two sub-sections outline island tourism management and island tourism research to support the management principles and practices that can improve the fortunes of island tourism.

1.2 Island Tourism Management

Island tourism management involves the management of tourism activities within island destinations. An island destination is defined as an area with amalgamated tourism products and services that provide a tourism experience within an island environment. Within any one country there can be one to several destinations. Hotels and attractions are the main co-located activities that would form the basis for destination designation. In identifying the location of a tourist destination, the transport element is a supporting resource to convey the tourist, and is not a prime influencer in increasing the tourist length of stay as would be hotels and attractions.

The scope and scale of tourism are important considerations of island tourism management. Scope relates to the range of tourism activities such as the types of hotels, attractions and mix of activities. Scale relates to the volume of tourism activity with respect to time and space. Seasonal variations relating to peak and non-peak times of tourist arrivals from the country of origin are the main elements of scale. Scope and scale combine to create a tourism pattern within island destinations. Understanding this pattern is a key ingredient in planning and implementing tourism growth and development strategies.

Some assessment of an island destination has to be conducted to depict the pattern of tourism activities and to apply the principles and practices of tourism management. Tourism management is the planning, organizing and controlling of tourism products and services to achieve some purpose or goal. As was mentioned previously, tourism activities create a pattern within island environments. This pattern has to fit within an existing physical space that is already inhabited and in some instances already constrained. It is an understanding of the enablers of the island tourism phenomenon that helps tourism management to create the greatest opportunity. With constrained resources, economic, sociocultural and environmental impacts and foreign dominance among other challenges, a real opportunity that this volume provides is to improve on the management of tourism within island environments. One of the major challenges for small island destinations in this century is the balance between sameness and uniqueness. Sameness is grounded in the need for efficiency, and therefore scale, in order to face rising costs in generating a tourist product; while uniqueness is related to the need for insular authenticity (Croes, 2011). Such an approach is supported by sound research approaches for policy making and problem solving and this will be addressed in the next section.

1.3 Island Tourism Research

New research perspectives and approaches are needed to improve the tourism practices within island environments, not only in the Caribbean, but across the globe. Traditionally, research about island destinations has focused on two main issues, sustainability and competitiveness (Graci and Dodds, 2010; Carlsen and Butler, 2011; Croes, 2011). In the island tourism research context, sustainability concerns the long-term viability of a tourism offer within the island destination whereas competitiveness concerns the resilience of the island destination to continually attract tourist markets. The issue is whether these two goals are mutually compatible, in that, competitiveness may not be sustainable, and sustainability may not lead to competitiveness. Hence the reason for the need examine and explore the island tourism phenomenon with new perspectives and develop new areas of research to refocus the dilemma between sustainability and competitiveness.

Island tourism research should involve a refocus on the main purpose for tourism activities within islands, which is for the host community to obtain benefit. This is a growing concern of island governments, businesses and academics (see Croes, 2016). Tourism activities should improve the quality of life of island inhabitants and provide profit for local communities. While tourists obtain a benefit through a rewarding visit to the destination, equally the host population must derive equal, or more than, gains from tourism activities. Refocus will provide a new perspective to obtain the right balance between hosts' and tourists' benefits, for win–win tourism growth and development, including addressing issues concerning access and profitability (see McLeod *et al.*, 2017).

From a demand perspective, emerging trends in the way tourism is bought and sold can result in a fall off of tourist markets. From a supply perspective, new types of tourism products and services, such as those in the sharing economy, have emerged to compete with traditional tourism offers. Dynamic changes in the business of tourism have an influence on the possibility of managing tourism within island destinations. In some instances, tourist arrivals have declined in the Caribbean (see McLeod *et al.*, 2017).

While tourism benefits in the short term seem clear, its benefits in the long run are not self-evident. Despite the significant impact of tourism globally, the resultant developmental effects on small island destinations have been uneven thus far. Issues and challenges, such as the impact of globalization on quality of life, competitiveness and uniqueness; the effects of climate change; accessibility; emerging demand exigencies triggered by demographic changes; Cuba's reinsertion into the Caribbean's tourism landscape and the potential impact on the competitiveness of Cuba and other Caribbean islands; and the constant alignment between the local industry and emerging demand patterns, require quick, savvy and smart destination management.

Given the dominance of tourism on warm-water islands with year-round tourism activities, it is important that a new course of research be charted around tourism management of island destinations. Warm-water island tourism destinations are a cohesive group of islands distinguished by geography and remoteness, history as former colonial territories, and dependence on external stakeholders for their economic and social development. Warm-water island tourism destinations also have a year-round tourism industry.

Tourism is a relational phenomenon and the activities and practices within the tourism industry are systemic. The island environment is contained in many instances over a small geographic scale and thus islands are ideally suited to study the systemic nature of tourism. Small island destinations are facing unprecedented adjustment challenges in the wake of the increasing process of globalization, and are in search of appropriate policy responses to that globalization. It is critical for small islands to understand how these challenges affect tourism performance and how they impact their residents.

1.4 Structure of the Book

This volume about tourism management in warm-water island destinations begins with this overview, Chapter 1, and is divided into three parts. The first part, Chapters 2 to 5, is about resources for islands, including transport and hospitality, and demand for islands. This part addresses the core resources of tourism products and services to suggest principles and practices to manage these in island environments. Chapter 4 addresses the demand for islands including the re-emergence of Cuba on the tourism landscape of the Caribbean. The second part, Chapters 6 to 8, concerns tourism policy, planning and development to guide appropriate forms of tourism development that will create benefit for islanders. The third part, Chapters 9 to 11, regards tourism marketing and management's role in exploring new strategies that can be adopted for successful island tourism management. The volume also includes a review of the network methodological approach in Chapter 12 for understanding island tourism and suggests that a new approach is needed to develop island tourism research.

Chapter 1 addresses a framework for the book including the terms and concepts regarding warm-water island tourism, a rationale for the book and its structure. The chapter outlines island tourism management and island tourism research and suggests a refocus on issues of benefit, access and profitability as new areas of island tourism research.

Chapter 2 concerns island travel and transport. This chapter considers air-transportation in relation to islands and highlights the policies under which islands utilize travel and transport services that support tourism. Of particular consideration will be the various types of transport services and the nature of these services and how transport supports the economic viability of destinations.

Chapter 3 is about an altruistic behavioural change approach to sustainable tourism when funds are scarce. Small tourism businesses require new strategies to gain competitive advantage in a global business such as tourism, and in that regard business certification is seen as an important tool. Access to these programmes comes with a substantial cost to businesses and therefore this chapter addresses a new approach in obtaining the 'labels' that are seen as an important operating practice for a tourism business. This chapter explores an approach taken by a tourism enterprise in the Caribbean island of Grenada.

Chapter 4 regards tourist motivation and demand for islands. The chapter focuses on the various demand models that are suitable for islands that are smaller in size and whether specific determinants of demand, for example, economic versus psychological, need to be addressed to support island tourism. A review of the nature of demand for the Caribbean tourism product involving an emerging tourist destination such as Cuba has been included. In particular, the analysis considers Aruba, Bahamas, Barbados, Jamaica, the Dominican Republic and Cuba.

Chapter 5 notes the changing face of Caribbean hospitality. This chapter considers the nature of hospitality services in islands and the changing types of tourists that warrant new approaches to ensure business success. New trends in consumer practices have been highlighted and the impacts on Caribbean tourism detailed. A suggested approach to provide quality tourism experiences has been outlined and a way forward has been charted to improve Caribbean hospitality.

Chapter 6 regards policy making for tourism development in islands by addressing policy-making networks to understand and improve tourism policy formulation. The chapter considers the actors involved in the tourism policy making process in Jamaica and Trinidad and Tobago to determine the characteristics of tourism policy making to meet the strategic tourism goals of both countries.

Chapter 7 concerns a re-emerging tourist destination by exploring planning for tourism growth in islands, a case study of Cuba. Various forms of tourism exist within island destinations; however, the creation of tourism destination success comes with astute plans for tourism development. The chapter outlines appropriate planning and development strategies for island environments and discusses the issues and challenges with these to suggest paths for tourism development success. The case of Cuba has been considered, regarding its transition and plans for tourism, as the country re-emerges within the region's tourism landscape.

Chapter 8 is about subjective well-being and tourism development in small island destinations. This chapter explores the literature pertaining to the relationship between tourism development and well-being from the residents' perspective. The relationship between these two constructs is relevant in understanding the role of tourism development in unfolding social dynamics in small islands as well as its impact on productivity and the economic prospects of these islands. The chapter assesses a number of topics addressed in the tourism literature. For example, the chapter examines if higher incomes provoked by tourism development make residents in small island destinations happier. Additionally, the chapter looks into the role of tourism development in enhancing social comparison as a force in defining happiness of residents in small island destinations. The gaps in the literature were highlighted and future research stemming from the insights gained from the literature review as well as directions for policy-makers have been discussed.

Chapter 9 explores tourism in the Seychelles using a consumer's perspective. Sound island destination marketing and management principles and practices are supported by knowledge of consumers' perceptions of the destination. The Seychelles, a small, remote island within the Indian Ocean, is heavily dependent on tourism for economic growth. In that regard, consideration should be given to the needs of tourists and locals alike to achieve a successful path for island tourism marketing and management. The chapter considers consumers' perceptions of the island destination using the netnographic technique to provide a basis for island tourism success.

Chapter 10 concerns a case study of destination management including crisis management of the Grand Bahama Island in the Bahamas. The chapter highlights the importance of appropriate tourism marketing and management strategies including the appropriate agency to function as the Destination Management Organization based on an assessment of network characteristics of density and centrality. The case study highlights the character of a destination in decline and suggests an approach to avert such a decline.

Chapter 11 explores the contribution of music festivals to tourism in small island destinations – in particular, in Aruba and Curacao. The chapter reviews the literature discussing the ability of music festivals to contribute to tourism development and economic growth by increasing the number of international tourist arrivals to small island destinations. In so doing, the chapter assesses the theoretical frameworks grounding the role and contribution of music festivals, and the empirical results stemming from these studies. The chapter discusses the gaps in the literature, followed by lessons learned from the literature in terms of promises and pitfalls of music festivals as an experience offering that can be used by small island destinations to enhance their brand equity and international tourist demand. Finally, the chapter provides some directions for future research and implications for destination managers.

Chapter 12 is about methodologies in island tourism research. This chapter considers quantitative and qualitative methodologies that are relevant to island tourism research. The chapter focuses on the network methodology and suggests a new approach to understand changes in island environments with dominant tourism products and services using social network analysis (SNA).

1.5 Summary

This volume highlights the tourism phenomenon in island environments for the main purpose of improving the benefits of tourism using a tourism management approach. A tourism management approach is utilized, supported by research that can highlight the existence of this unique form of tourism. The need to develop appropriate principles and practices of tourism management that can result in island tourism success has been outlined in the structure of the book with a review of methodological approaches that can contribute to the future success of island tourism.

References

Apostolopoulos, Y. and Gayle, D.J. (2002) *Island Tourism and Sustainable Development: Caribbean, Pacific, and Mediterranean Experiences*. Praeger Publishers, Westport, Connecticut.

Boniface, B., Cooper, C. and Cooper, R. (2012) *Worldwide Destinations, the Geography of Travel and Tourism*. Routledge, Abingdon, UK.

Carlsen, J. and Butler, R. (2011) *Island Tourism: Sustainable Perspectives*. CAB International, Wallingford, UK.

Central Intelligence Agency (2017) The World Factbook, Physical World map. Available at: https://www.cia.gov/library/publications/the-world-factbook/graphics/ref_maps/physical/pdf/world.pdf (accessed 26 March 2017).

Croes, R. (2011) *The Small Island Paradox. Tourism Specialization as a Potential Solution*. Lambert Academic Publishing, Saarbrucken, Germany.

Croes, R. (2016) Connecting tourism development with small island destinations and with the well-being of the island residents. *Journal of Destination Marketing & Management* 5, 1–5.

Graci, S. and Dodds, R. (2010) *Sustainable Tourism in Island Destinations*. Earthscan, London.

Holder, J.S. (2013) *Caribbean Tourism*. Canoe Press, Jamaica.

McLeod, M. (2015) Caribbean Tourism. *Tourism Management* 48, 316–317.

McLeod, M., Lewis, E. and Spencer, A. (2017) Re-inventing, revolutionizing and transforming Caribbean tourism: Multi-country regional institutions and a research agenda. *Journal of Destination Marketing & Management* 6 (2017), 1–4.

PART I: Island Tourism Transport and Hospitality

2 Island Travel Transportation

VINCENT VANDERPOOL-WALLACE*

Bedford Baker Group, The Bahamas

'We are like islands in the sea, separate on the surface but connected in the deep.'
William James

2.1 Background

Every island is surrounded by a moat. In times of war, it is a wonderful natural defensive barrier. For islands dependent on tourism for economic development the corollary is also true. Islands are difficult to supply with goods, services and visitors, and with the increasing width of the moat comes increasing difficulties with supply. The width of the moat matters because if every island were close to another, or to a large mainland, bridges would be built across the divide to facilitate access. With the increasing width of the moat, ferry vessels, then larger ships and aircraft, are soon required for passenger transportation and supplies. It is self-evident that bridges facilitate commerce. It is self-evident that moated castles need drawbridges in order to receive supplies. It is self-evident that inexpensive meals and shows facilitate the attraction of patrons to casino floors. It is becoming increasingly self-evident that for tourism-dependent islands, airlines are merely air bridges to facilitate the commerce of tourism. All of these difficulties are solvable but they all make managing island tourism a special case requiring special considerations.

Governments that are owners of bridges do not ask about the profitability of a bridge because, to them, and to their citizens, the bridge's value is self-evident, but as soon as government-owned or government-subsidized ferry vessels, aircraft and ships are introduced to the conversation for connecting to islands, the conversation changes immediately to the profitability of these forms of transportation. No one would ever recommend the demolition of a bridge on the basis that its tolls cannot

*E-mail: vVw@bedfordbaker.com

cover the cost of construction and operation. The value of a bridge rests on its value in facilitating commerce, the obvious reason for building the bridge in the first place. Moreover, in order to ensure that the facilitation of commerce is not impeded in any way, the bridge tolls, if any, are kept to a minimum. The revenue for the payment of the bridge comes from the taxes on the commerce that it facilitates.

The same kind of thinking is what drives casino companies to surround a casino floor with showrooms and restaurants. In many cases the price of restaurant meals and the price of the shows are kept deliberately low because they are bridges to bringing patrons to the casino floor. Governments and private sector practitioners on tourism-dependent islands could be guided by the management strategies of bridge builders and casino operators.

Airlines and ships serving tourism-dependent islands are nothing more than air bridges and sea bridges and they need to be regarded in the same way as other bridges. What matters to the owners of casinos is the volume of play on the casino floor created by the attractive food, beverage and entertainment offers around the casino. The attractive food, beverage and entertainment offers are paid for by the increased 'casino action'. What matters to the builders of physical bridges is the commerce created and facilitated on the island to which the bridge is connected. The bridge is paid for by taxes on the increased commerce stimulated by the bridge. What should matter to island tourism managers is the increased tourism created and facilitated on the island that the airlines and ships serve. The prices of airline seats and the price of ship cabins to the traveller should be kept to a minimum by the collection of taxes on the incremental tourism created on the island.

When the marginal cost of keeping food, beverage and entertainment prices low equals the marginal benefits to the casino operation, equilibrium will have been reached. Physical bridges are widened or extra bridges are built and tolls are kept low until the value of commerce on the island is optimized. Low-price air transportation and low-price sea transportation to any island should continue to be supported until the tourism value on the island is optimized.

2.2 Tourism Necessitates Travel

The English word 'travel' is believed to be derived from the Old French word 'travail' meaning labour or toil, which migrated into English and was an apt description of the arduous experiences involved in moving from place to place for an increasing number of people in 14th century England. It appears that only after wayfarers found the journeys less arduous and those promoting the journeys amplified the delights awaiting at the destination that the etymological change was made to 'travel'.

Indeed it was improved and more comfortable transportation that made way for today's focus on the hospitality and experiences at the destination. As early as the 14th and 15th centuries, words such as 'recreation' from 'recreate', 'vacation' from 'vacate' and 'holiday' from 'Holy day' began to emerge. Curiously enough, the words 'hospitality', 'hospital', 'hostel' and 'hotel' are all derived from the same Latin word 'hospitalitas' meaning 'friendliness to guests'. So a 'hospital' was originally intended to be a place where one could be friendly and comforting to the ill.

At its core, leisure travel today is still driven by the expectation of hospitable hosting and caring for guests all along the chain of custody from the time they leave their place of residence to their return. Generally speaking, though important, travel today is much more focused on the joys of the destination and much less focused on any discomforts of the journey, which have been much diminished.

Islands as destinations have a special lure for travellers. Many today believe that there is some visceral connection that human beings have to islands as places to visit. When one adds year-round warm weather to an island destination it becomes naturally attractive to people nearer the poles especially during their winters. This is evidenced by the retention of the phrase 'the winter season' to represent the period of peak demand. In addition, the relatively recent increase in the desire for leisure travel to warm-weather islands is derived not only from the visceral and instinctive human yearning for islands, but also from the decreasing travails endured in travelling to those islands because of significant improvements in transportation.

It was perhaps the Greek islands in the last 100 years that really began to take advantage of better and easier transportation. Those early promoters then layered on top of the better transport systems the idyllic and paradisiacal references associated with many islands in literature. Once the issues with access were overcome, prospective travellers began to focus on the joys at the end of the trip. In fact, even in the case of the Greek islands with their deep and well-regarded history and historical sites, much of the leisure travel was driven primarily by the enjoyable weather available on the islands with the history acting as a 'tie breaker' against competing destinations in the traveller's consideration set.

Today, cruise companies offering stopovers to pristine, unpopulated islands find that those are the most highly rated experiences. It appears that for the majority of travellers, the warm natural pure and uncontaminated islands are indeed strong attractions as long as one can repair to the comforts of modern conveniences when desired. While the Robinson Crusoe fantasy is universal, the amount of time one desires to be spent away from modern conveniences is relatively brief, except for a very small group seeking adventure travel.

So it was not surprising that a study carried out by MasterCard found that islands, and in particular warm-weather islands, appear to be the most desired destinations for global travellers (MasterCard Advisors, 2006). Similarly, an annual study by MMGY Global, an international marketing company, showed that over a 36-year period from the company's inception until 2017, the Caribbean was regarded as the most desired foreign destination year upon year for the majority of US citizens.

In fact, much of the desire for travel to warm-weather islands began before the age of air travel. In the late 19th and early 20th centuries, those who could afford it travelled thousands of miles under relatively uncomfortable conditions as compared to today's modes of travel to get to warm-weather islands that had earned a reputation for their healing, restorative and salutary powers. There is little question that the most significant advance in island tourism was ushered in not just by air travel but specifically by the invention of the jet plane. Without that invention the economic contribution derived from tourism for so many island destinations would be decimated. It is now widely established that islands, especially warm-weather islands are highly desired as leisure travel destinations globally. The natural desire and demand for warm-weather islands is moderated to a significant degree by transportation constraints.

The improvements that ushered in the golden age of leisure travel began as much with ships that were far better suited for long distance travel as roadways were then notoriously difficult to traverse. From the perspective of island destinations, however, except for ferry transportation, ships are relatively slow, require deep water access and deep water ports. That is why the early ocean-going vessels often required ferrying of passengers and goods to and from the ships to the port because of unnavigable reefs and shallow water. In earlier times, the length of the voyages to islands made them accessible only to a very few. Even today, especially for multi-destination cruises, the need to commit to the full cruise and the relatively short period of time spent in each port makes cruise passengers an important but economically inferior contributor to tourism value.

In terms of road transport, macadamized roads beginning in the mid-1800s, and later tarmac (for tarmacadam) in the early 1920s, made road transportation far more comfortable and efficient. Except for some rare cases such as the 160-mile drive from Miami to Key West, Florida over 42 bridges, road access to distant islands is unavailable.

2.3 The Practicalities of Tourism

Tourism is that part of an island's GDP that is derived from the economic activities of visitors. It is a difficult economic sector to grasp without special economic tools that will be discussed later. An easier and simpler way to assess tourism is to regard it as a daily growth of the population of an island by visitors who have a high propensity to spend, and in many cases the average daily expenditure of the visitors is higher than that of the most upscale community in the island. While the destination grows its numbers of high-spending visitors, it does not have to increase the stock of roads, hospitals, schools and other costly services that are normally provided for the permanent residents. Furthermore, for many island destinations, the spending comes in the form of very valuable foreign exchange. So, by and large, tourism brings a steady stream of incremental sources of hard currency into the island economy with relatively low costs to the society. Visitors to many island tourism destinations may be regarded as individual foreign direct investors seeking only positive memorable experiences for their investment, not assets.

That is what makes tourism a valuable economic development tool that has proven to be of substantial value to many island destinations often representing up to 80% of their GDP. Few countries would feel comfortable with such a large reliance on one economic sector so it is not surprising that the higher the GDP contribution derived from tourism, the louder the cry for economic diversification. Furthermore, it does not help matters that in many island destinations, a second pillar of their economies and a source of business traffic comes from their financial services sector, which is under severe attack from OECD countries. Instead of economic diversification increasing in these tourism-dependent countries, it is diminishing. The recent fall in oil prices and its effect on oil-dependent economies has not gone unnoticed by governments in these highly tourism-dependent economies. The fear is that what has happened to oil-based economies, can also happen to one so highly dependent on tourism.

What is often missed in many of these tourism-dependent islands is the relatively small size of the population, which can be as few as the 55,572 residents on St. Kitts and Nevis in the Caribbean, for example. Diversification, while desirable, is most difficult with these very small populations (World Bank, 2015). As with oil-dependent economies, these island countries did not choose tourism as their principal industry, tourism chose them. They discovered from the spontaneous increases in visitors that they had a comparative advantage in tourism and thereafter chose to promote and manage it deliberately by establishing ministries and tourism boards. There was no master plan for national development with governments selecting pursuit of tourism development from among a number of possibilities. So with tourism, island governments would be wise to 'make hay while the sun shines' even while they search for other opportunities for which they have the competitive and comparative advantages to satisfy the desire for economic diversification, but there is always more regarding the logical factors affecting tourism.

There is still the need to address the powerful sociological and historical factors that affect tourism. One such factor is the extant, though declining, belief by many that any global demand for leisure trips is necessarily fickle and unreliable. People have to eat; people don't have to travel for leisure. That is held as an article of faith by many national policy makers. That explains the exasperating desire to cling to agriculture as a primary industry on many islands; and despite the budget allocations for tourism, the otherwise tepid support that 'serious' people give to tourism. No matter how many examples are brought forward in support of the sustainability and resilience of island tourism, many remain impervious to the evidence.

In addition and most unfortunately, many of the island destinations that are now highly dependent on tourism suffered through the legacy of slavery (see Holder, 2013). So there is the extant, though declining, view that tourism is a remnant of the sons and daughters of former slaves serving the sons and daughters of former slave holders. As a result, even though tourism is often the largest sector of the island economy, many of the best and brightest of the island's citizens shun involvement in tourism in favour of other professions in medicine, law, engineering, accounting and the like that deliver higher social status. A part of the difficulty in tackling the negative perception of tourism is the unfortunate fact that some of the more visible positions in tourism are those with the lowest earnings, viewed as requiring relatively low levels of academic achievement.

This lingering perception explains, in part, the lack of rigour in much tourism analysis and the persistence in viewing tourism in terms of the oft used phrase: 'It's not rocket science'. There is a lingering colonial overhang and general suspicion in many island destinations about the centrality of tourism as a platform for national economic development.

In reality, tourism falls in the penumbra of economic science that makes it one of the more complex sectors to describe. It is impossible to get one's arms around the sector and measure its impact accurately and manage it properly without such sophisticated rocket science tools as 'tourism satellite accounts' (TSA). So the sector is confounded by this conundrum in that it requires rocket scientists to plan, manage, understand and promote it but it is still regarded by most as being simple and fickle. Tourism is a complex web of services that is best unravelled and understood with the use of TSAs. In the same way that a series of tests is needed to determine human health, a series of determinations under the TSA is required to diagnose and determine an island's tourism health.

No bank would dare tell its shareholders the number of people who came through their doors was a measure of their monthly or annual performance. The number of vehicles crossing a bridge cannot be a measure of the value of the bridge. Similarly, tourism destinations cannot rely solely or even principally on the number of visitors arriving by air as a measure of the health of their tourism economy. Yet, that has been standard practice for decades.

Governments of every island in the world spend considerable amounts of time seeking foreign and local direct investments. Tourism is the path by which they receive those direct investments from individual visitors without having to give up any physical assets in most cases. All that visitors require are positive memorable experiences. Which would an island destination prefer: 100 visitors staying for one night or 100 visitors staying for one week? In each case the island will record 100 visitors but in the second case, in addition to the expenditure for nightly accommodations, the second group will consume more meals and likely spend more on island transportation services at a minimum. Which would an island destination prefer: high-quality tomatoes that are grown locally and consumed by both residents and visitors or imported tomatoes of the same quality for the same prices? Which would an island destination prefer: tourism entrepreneurs and workers who repatriate half of their income to their homeland or tourism entrepreneurs and workers that are all local? These and many others are questions that can be tracked with a TSA to deliver the kind of knowledge to island tourism managers that would enable the optimization of tourism benefits.

2.4 Transportation and Tourism on Islands

The value of tourism does not begin for islands until and unless the visitor arrives at his or her destination. The central point of this chapter is that, for island tourism, it is as foolhardy to discuss island tourism without a thorough understanding of and some responsibility for managing the air transportation and sea transportation that deliver visitors to the island. Responsibility for transportation to islands and responsibility for tourism development on islands must be inextricably intertwined. That will soon become self-evident.

In landlocked areas, most visitors arrive from outside the area by roadways. For France, the largest leisure destination in the world, the overwhelming majority of visitors arrive by roadway and railway with most of them coming from the 508 million surrounding population of European Union countries (European Union, 2017). Most people are surprised to find that for Las Vegas and Orlando, the two largest leisure destinations in the US, most visitors arrive by personal vehicles whether rented or owned. In the case of Orlando, 50% of its tourism comes from the surrounding state of Florida. In the case of Las Vegas, 25% of its visitors come from the adjacent state of California, the largest state by population in the US.

Furthermore, unlike air transportation to islands, there is no competition for seats in personal vehicles between residents and visitors. The competition for airline seats between residents and visitors affects both the availability and prices of those seats to the island destination. Further still, unlike scheduled air transportation to islands, personal vehicles move on roadways at their own schedule and pace to their destinations. The flexibility provided by roadway access from distant cities has a substantial impact on property occupancies and yields at the destination. In fact, that flexibility is what makes some initiatives for landlocked destinations relatively easy and makes those very same initiatives for island destinations impossible.

For example, one persistent and emotional shibboleth among many islanders that is often supported strongly by the local population and local press is the desire to attract very large groups of international visitors to fill local stadiums for such events as football games, cricket matches or music concerts. For island tourism managers, the largest impediment to the success of these events is not marketing but arithmetic.

Any simple simulation will show that with the available supply of daily airline seats, the seat prices and seat supply quickly become impossibly prohibitive to the success of the event. For those who suggest the booking of charter airlines, they soon find out that the cost of aircraft waiting for the group or the cost of flying two legs of the flight empty to collect patrons at the end of a multi-day event is similarly prohibitive. In addition, because of the start and end dates for the event and the need to move very large numbers of patrons on a limited number of days, occupancies of properties are so distorted that their effects on monthly revenues of the properties are often not worth the effort. For example, in order for island properties to make rooms available for a very large check-in, they must artificially run rooms vacant for several days and when there is a very large check out of the group, it is impossible to fill the large number of vacant rooms immediately because of the constraint of available scheduled seats. Island managers soon learn that their best strategy is to seek out groups and individual travellers that can be accommodated year-round with the existing supply of scheduled seats and to work very hard to avoid large peaks and valleys in demand.

Again, for landlocked locations where most patrons arrive and depart with their own transportation and often without the need for accommodation, support for large events is far easier. That is why for islands, much of the attendance for large events must come from the local population, *not* from visitors. In many cases, for a variety of reasons including the cost of accommodation, it helps considerably if a significant proportion of the visitors are accommodated by friends and relatives as is the case for many annual music events on island destinations.

The forgoing also reveals a psychological disadvantage that tourism managers on islands must also endure. Canadians, for example, will drive from Toronto to Baltimore to board a cruise ship heading to the Caribbean and they will not regard the time and cost of their drive to the port as a part of their total vacation cost especially when using their personal vehicle. Those same Canadians flying by air from Toronto to the Caribbean will see the cost of airfare as a part of their total vacation cost and do all that they can to secure a lower priced airfare.

Case study 1

The Cricket World Cup in the Caribbean in 2007 was a classic example of the constraints imposed on large events by the supply of scheduled air transportation. One enthusiastic destination, expecting an additional 10,000 visitors for their cricket match encouraged their country's residents to borrow funds, before the advent of easy peer-to-peer systems such as Airbnb, to extend their homes to accommodate the needs of the incremental demand for accommodation that was forecasted. Setting aside the wisdom of borrowing funds for the long term to accommodate a very short-term opportunity, tourism managers apparently forgot that 10,000 visitors would require the equivalent of an additional 40 full Boeing 757s flights to their countries over a 1-week period. The country then turned to a cruise line to provide the transportation and accommodation for the expected demand only to find that the line required the country to not only cover the cost to get the ship to the port, but also the opportunity cost of taking that ship offline for the requested time.

Let us assume that the fans arrive over a 5-day period prior to the game and on average stay for 7 days. That would require eight additional 757s per day and those early arrivals would have to pay daily sums for meals and hotel accommodation for their 6-night stay all for the enjoyment of, at most, 2 days of matches. Then the air transportation providers need to recover an additional charge, which must be absorbed by someone. Since the flights are incremental, they will either have to wait on the ground for the 7 days or make the round trip to their home airport empty in order to collect their charter passengers. Suddenly the price of attracting those 10,000 incremental visitors becomes prohibitive.

By any measure, the Cricket World Cup in the Caribbean underperformed the promises made by the promotors simply and largely because it was both logistically impossible for island air transportation to accommodate their ambition and financially prohibitive for fans to travel to the destinations for the events (see Jordan, 2006; Jordan, Tyson, Hayle and Truly, 2016).

If the Cricket World Cup were held in a destination that allowed for a large number of fans to arrive by land, the situation would be much different. Those arriving with their own or rented vehicles can stay for however long they wish and in many cases they drive in just for the game and return home. Such an option is often not available for islands. Compare, for example, the average cost of an English Premiere League Football match of £65 according to a report by the Campaign for Better Transport (2013). The £65 includes transportation, food and drink and the match ticket. For the Cricket World Cup, the average costs to attend a match in an island destination for many patrons would be in the thousands.

Case Study 2

The second case involves an island destination that determined that it would be of great economic and publicity value to attract the 2000-room double occupancy International Monetary Fund (IMF) conference, which needless to say would wish to come during the peak month of the year for their 4-night stay. Since the capacity of airline seats to the destination is approximately 5000 daily and because many of those seats are booked for vacationers already in the peak season, only 50% of the capacity might be available for the 4000 persons attending this conference. So, in essence, over the 2 days of their arrival, the conference attendees will consume all of the remaining seats from the various markets from which they arrive.

The availability of seats was not the problem. The problem was the need for the hotels to so manage their room occupancies that 1000 rooms would be available on each day of arrival for the group. The only way to manage the check-in for such a group is to allow an abnormally large number of rooms to remain vacant prior to the arrival of the group in order to accommodate the very large check in of delegates, which would lower both the average occupancy and RevPAR for the property for that peak month when compared with prior years.

The reverse happens on departure when the properties return slowly to normal occupancy with such large checkouts. At the end of the month the destination and the properties would find themselves far worse off for the experience. Incidentally, the airlines also suffer under such a scenario because properties that cannot accept reservations are reflected in empty airline seats at a peak earning period. Fortunately, in the case of the IMF conference, logic prevailed; the group was persuaded to travel to another destination and the island in question that booked the group originally had its best month ever.

Again, in destinations with access to a large volume of personal vehicles, it is possible to promote and sell a number of short stay visits leading up to large check-ins and check-outs. Such options are largely unavailable to island destinations.

2.5 Bridging Island Transportation

To reiterate, if islands were closer to each other or to the mainland, connecting bridges would be built to facilitate tourism like the 42 bridges traversed to travel the 160 miles from Miami to Key West Florida. Caribbean islands, for example, receive 0% of their hotel visitors by ground transportation and 100% by air, since island destinations are surrounded by water. What is more, there is never a discussion by anyone about the profitability of the 42 bridges that enable ground transportation to Key West and the total tolls are under $7.00 per vehicle. Nevertheless, as soon as a tourism-dependent island destination has to guarantee airlift to secure seats for visitors or for the government-owned airline to provide transportation, the conversation immediately pivots to the profitability of the transportation.

All of the foregoing reinforces the fact that for most tourism-dependent islands, air transportation and tourism are so inextricably intertwined that responsibility for tourism development and responsibility for air transportation should be tightly integrated. Tourism benefits do not begin to flow to an island until and unless a visitor arrives. Any island tourism manager that is not also directly engaged in the discussion to provide low-cost, high-frequency, high-quality non-stop air transportation is a severely handicapped tourism manager.

It is fully recognized that island tourism can also rely to some degree on visitors arriving by cruise ship. Even though the expenditures by cruise passengers for accommodation, for transportation and for some purchases remain largely with the cruise company and the expenditure on the island is relatively small, those on-island expenditures are highly concentrated and go to a segment of the island tourism economy that is highly dependent on those passengers. Yes, the cruise lines also suggest that each destination should seek to get the cruise passenger to come back for a hotel stay. Any analysis of those opportunities show that this is brilliantly palliative and a case of misdirection by cruise companies to take focus away from some of the irksome issues that persist with regard to their operations. Even if those issues were resolved, air visitors are by far the more valuable contributors to the tourism economies of islands.

Technology today has revolutionized the world of travel. Nowadays, air transportation is so easily accessed that we take it for granted. It is a necessary means to an end. Transportation can only dissatisfy the traveller if it is inconvenient or prohibitively expensive. That is why, other factors being equal, island destinations will always attract more visitors from nearby source markets compared with faraway sources.

Nearby destinations are expected to be lower in transportation cost and more easily accessible to the traveller. There is also the factor that island tourism managers frequently forget: the further away the source market, the higher the airfare, the longer the length of stay and therefore the higher the total cost for the traveller.

That is why many pronouncements by island tourism managers about 'going after' large numbers of visitors from distant and opposite sides of the earth are largely exercises in self-flagellation. From the traveller's point of view, such travel requires ignoring similar destinations nearby, jet lag adjustments, much higher airfares, often multiple layovers and a necessarily long length of stay at the destination. The cost and discomfort of this exercise diminishes the number of prospects in the target group to a trickle. Unfortunately, as in times of yore, the more difficult the passage for the traveller the less likely they are to choose the destination.

Fortunately, the converse is also true for prospective visitors from nearby source markets. Since the cost of access is relatively low and because they are generally in the same time zone and because the traveller can control their length of stay, the target audience for smart island tourism managers will vary according to the source market of the visitor. It is entirely possible for a middle income traveller on an overnight stay from a nearby destination to exhibit the spending pattern of a wealthy patrician from a much more distant source market. Perceptive island tourism managers quickly discover that unlike commodities where the price of a bottle of Ketchup in a Walmart in Miami is often the same as in a Walmart in Seattle, the price of a vacation to the nearby Bahamas from Miami can be a fraction of the cost of a vacation to the Bahamas from Seattle.

Island tourism has the peculiarity of having only one outlet on Earth and the customer has to travel to the outlet to enjoy the product. It stands to reason that the cost of getting to the outlet then becomes a part of the product and unfortunately in the case of tourism, the unit cost of transportation is larger than the transport of any other product.

So for island tourism, proximity matters and it matters a great deal. To further the example just provided, for all of the reasons cited, there is a far, far higher likelihood of the visitor from Miami traveling to the Bahamas multiple times each year compared with the traveller from Seattle. So there is little wonder that the State of Florida and the eastern seaboard has been the largest source market for the Bahamas and most of the northern islands of the Caribbean for the past four decades. The same pattern is seen for the southern Caribbean islands of Aruba and Curacao. A substantial number of their visitors come from nearby Venezuela; therefore proximity matters.

2.6 Price Matters

Distance and multiple stops are dissatisfiers for air travel trips to island destinations. So is the price of travel. In fact, many travellers will trade off a non-stop flight for a lower cost flight with one or more stops. It stands to reason that low price non-stop flights are most desired and that travel is stimulated by the introduction of low price non-stop flights to destinations.

This finding about the stimulative value of airfare is now beyond dispute and is corroborated by studies of the so-called 'RyanAir effect', 'JetBlue effect' and

'Southwest effect'. Whenever these airlines enter the market their lower fares alone result in measurable increases in visits to the destination to which they fly. From the traveller's perspective, it matters not what policies deliver low-price airline tickets. Low ticket prices are all that matters in stimulating air travel. Even though Las Vegas and Orlando receive more visitors by ground than by air, the two destinations are perennially among those with the lowest airfares in the United States in terms of price per mile flown. They clearly understand and benefit from the value of low-cost transportation in maximizing their tourism benefits.

The price of an airline ticket is comprised of four general components: (i) the price that the airline expects to receive for their seat; (ii) the taxes imposed by the government; (iii) the fees associated with the capital cost of the airport facility; and (iv) the fees attached to some ongoing operation of the airport facility such as security and similar services.

In many cases the sum of taxes and fees added to airline tickets to island destinations are higher than the cost of the airfare received by the airline. It is the total cost of the air ticket: airfare + taxes + fees, to which the customer reacts. Yet, whenever airlines conduct their analyses of demand for any island destination, it is nearly always examined from the perspective of their revenues alone and *not* from the perspective of the total ticket price as presented to and as paid by the customer.

In many cases the full ticket price being paid by the customer is easily equal to or twice the sum received by the airline for its services. The airline analyses mask the significant reduction in demand that tourism-dependent islands cause with their substantial additions of taxes and fees.

One destination has tested the effects on demand of removing taxes and fees from airline tickets and found a significant spike in demand. This was done by presenting the traveller with the base airline cost and the Department of Tourism paid the applicable taxes and fees, which was the least complicated method of testing the effects of lowering airfare prices. That same destination and others have examined the effects on demand of offering travellers 'companion fly free' offers and rebates on airfare equivalent to the cost of taxes and found it to be a significant stimulus. One very large resort at the destination has shared data to show how conversion rates more than triple after the rebates on airline tickets are added to the promotion without any other changes in the holiday price. It is now inarguable that the stimulative value accelerates with the size of the reduction in the cost of the airline ticket as paid by the customer.

The finding of equal importance was that the increase in demand was broadly felt across all types and categories of accommodation. This outcome can be viewed as a crude sensitivity analysis. It appears that reducing the cost of airfare stimulates travel to the island destination to a much greater degree than reducing any other prices such as room rates and food and beverage prices. This finding is very much in line with other studies that appear to suggest that leisure travellers buy the anticipated destination experiences and their accommodation is primarily their affordable base camp from which to launch their forays to enjoy those experiences. Airfare is viewed primarily a means to an end, a commodity. So any reductions in the cost of that commodity leave more funds available for a more comfortable base camp and more destination forays. That discovery is music to the ears of island destination managers because in many places, room rates are the most profitable and most important economic component of the vacation expenditure.

When one recalls the value of proximity in driving demand for an island destination, it is baffling to find so many tourism-dependent island destinations exacerbating matters with the legacy habit of adding more and more high fixed taxes and fees to airline tickets. That practice disproportionately dampens demand for the very nearby markets from which they should have the strongest demand. To wit, if the airline charges $100 for its airfare and the destinations add a $100 concoction of fixed taxes and fees, the total ticket price would have been increased by 100%. If the base airfare from a faraway country is $1000, however, the total ticket price would have been increased by only 10% with the addition of the $100 in fixed taxes and fees. Given the power and value of proximity in driving demand, fixed taxes and fees have the effect of disproportionately reducing demand for the very best prospects.

2.7 Regional Airlines

Regional airlines fail for precisely the same reason that monopoly or near monopoly national airlines fail in their attempts to provide quality service in a multi-island archipelago. Bahamasair the national flag carrier of the Bahamas provides a good example.

Whenever monopolies provide water, the quality and the pressure might be poor but at a minimum, generally speaking everyone gets water when they need it. Whenever a monopoly provides electricity, the price and the reliability might be poor but at least everyone gets electricity for most of the time whenever they need it. Whenever a monopoly provided telephone services, the price and the reliability might be poor but at least everyone gets communication services whenever they need it.

Since an aircraft is a discrete entity unlike continuous flow products such as electricity, water or telephone services, it is impossible for a small single-airline monopoly to provide everyone with services at the time that they prefer to have it. As in the Bahamas, most Bahamian business travellers on every island wish to leave their island early in the morning and return late in the evening. This is impossible to provide with a limited number of aircraft. In addition, the volume of traffic between any two desirable points varies considerably so the size of aircraft required will vary considerably. If all of the aircraft of the monopoly provider is mostly the same size as is often the case, there is very poor matching between supply and demand.

The solution in the Bahamas has been the enabling of private airlines to provide service alongside the once monopolistic national flag carrier. There is better matching thereby with supply and demand. Frankly, the next step in the provision of these services would be for the disappearance of the national flag carrier from competing with private airlines with the government looking to subsidize uneconomic routes instead of subsidizing an entire airline. The government could then also focus on securing efficient management services for airports, ensuring the quality maintenance of private airline services and even ensuring an efficient and integrated reservation service.

The same applies to the regional airline, LIAT, which is owned by multiple Caribbean governments and is a noble attempt to solve the region's air transportation problem, however, its current and past structures are instead impediments to regional growth in the Caribbean for all of the reasons described. It is far better to enable an open skies regime across the Caribbean along with the reduction is the cost of intra-Caribbean air tickets (see McLeod, Lewis and Spencer, 2017). There is now incontrovertible evidence that the increase in ticket prices as a direct result of the addition of bilateral taxes and fees has led to a reduction in intra-Caribbean passenger travel.

Considering all of the factors outlined in this chapter, it is very clear the largest potential leisure market for the Caribbean is the Caribbean itself, which now has the highest taxes per air mile flown of any region in the world. One supposes that some regions would be so adjudged but surely it should not be that region that is the most tourism dependent in the world. The bilateral removal or significant reduction of all airline taxes and fees should be an imperative.

Finally, while the world, led by the largest economies, was busy removing barriers to the movement of goods across borders in the interest of free trade, many of those same large economies went in the opposite direction with respect to the movement of people across borders. There has been a veritable explosion of increases in departure taxes, air passenger duties, security fees and passenger facility charges. It is very odd that world trade agreements have removed so many of the taxes on goods travelling across borders yet we have seen steady increases in tariffs on people crossing borders. After all, a departure tax is nothing more than a tariff on people and hurts those economies whose primary export is travel and tourism. Here is an example that best illustrates the matter.

A flight from New York City to Miami is only 9 miles shorter than a flight from New York city to Nassau, Bahamas, yet the combined taxes and fees on the tickets to the Bahamas is $130.00 higher per person. For the typical family of four, that additional $520 is a significant encouragement for that family to keep their trip within the US.

Tourism-dependent islands would do well to lobby the first world economies, especially, for the reciprocal removal of the departure tax in the interest of 'free trade'. Such an initiative is supported, in principle, by the United Nations World Tourism Organization (UNWTO), by the World Travel & Tourism Council (WTTC) and the World Tourism Organization (WTO). One explicit principle enshrined in WTO policies is that advanced nations should not have tariffs higher than developing countries for the same item. The departure tax applied by the US is nearly always higher than the departure tax of any of the countries of the Caribbean.

2.8 Closing Remarks and a Way Forward

So where are we now with regard to growing demand for tourism-dependent island destinations?

We know that island destinations are heavily disadvantaged by not having the option of receiving visitors by roadway, railway and the personal vehicles of their guests. We know that visitors arriving by air are by far the most important contributors to the tourism economy even though visitors arriving by cruise ships and other seagoing vessels can be very important to a small segment of the tourism-dependent economy. We know that, other factors being equal, those prospective travellers that are nearby are much more economically valuable compared with visitors from further away for a variety of reasons. We know that travel to a destination is more sensitive to changes in the price of airfare than changes in prices of any other factor. We know that the lower the price of airfare, the greater the stimulus for travel to the destination to all traveller demographics and to all categories of accommodations. We know that fixed taxes and fees when added to the base airfare of the carrier have the effect of disproportionately dampening demand from the very nearby markets that would be most important to the island destination. We also recognize the foolhardiness of Ministries of Finance adding taxes to tickets such that carriers demand that flights to the destination have to be guaranteed by the relevant Ministry of Tourism, which often uses its promotional funds for the guarantee thereby rendering them unable to promote the service. We also know that airline monopolies in a multi-island environment are inherently and extraordinarily inefficient in satisfying customer demands. We also know that advocates of free trade should tackle trade in travel for the reciprocal removal of departure taxes, especially since the current arrangement with developed countries having higher departure taxes than developing countries is a violation of WTO principles.

Policy makers at tourism-dependent island destinations have now come to appreciate the folly of failing to recognize their special circumstances. They recognize that airlines are the vital air bridges to their destinations and that air travel is facilitated to the degree that the 'toll' for the air bridge is as low as possible and the frequency of crossing is as high as possible.

The predicament is that many of these island destinations are so financially stretched that they are uncomfortable with the risk of reducing the price of air access by removing taxes and fees from airline tickets without some assurance that those taxes and fees would be returned with certainty from other economic activities. The task therefore is to significantly reduce or eliminate the risk associated with removal.

The Rt Hon. Perry Christie, Prime Minister of the Commonwealth of the Bahamas, in a speech to the Heads of the CARICOM group of countries in the Caribbean in 2015 said:

> I know that for some, tourism may be redolent of a part of our history that we would want to keep barricaded. For most of our countries, tourism is the largest earner of foreign exchange. For most of our countries, tourism is the largest employer. For all of our countries tourism absorbs the broadest range of skills of any economic sector. For all of our countries, tourism is the one sector for which there is no such thing as a jobless recovery. The very nature of tourism requires more people to be hired with increasing number of visitors. If unemployment especially youth unemployment is the scourge of our times, there appears to be no better economic sector for us to embrace in leading us closer to the promises that we have made to our lands (Caribbean Tourism Organization, 2015).

As a result of the many advances in technology, the global economy is growing but without the proportionate increase in employment leading to the relatively new economic phenomenon of 'a jobless recovery'. Tourism, because of its very heavy labour component for personal services, is among the few global economic sectors for which there is no such thing as 'a jobless recovery'. As occupancies increase there is a nearly direct increase in employment. Tourism also has the advantage as an economic sector because of the very broad range of skills that can be absorbed by the sector. Besides, any government would far prefer receiving their taxes from the activities of employed persons rather than from tax collections at the border, which has been shown in the PwC study for the UK to be a tourism job killer (PricewaterhouseCoopers, 2013).

As Prime Minister Christie notes, for tourism-dependent islands, the bane of their existence is high unemployment. So stimulating travel delivers in many cases, foreign exchange, increased linkages to indigenous products and services, a broader distribution of income and of greatest importance and increased employment. There is a direct connection between lowering the cost of access to islands, which leads to increasing the flow of visitors and increasing employment for island residents.

The exercise involved in lowering the cost of air access is simple. First, island destinations must determine how all of the current sum of taxes and fees now included in airline tickets can be covered by taxes applied to tourism-related and other transactions on island. Second, plan to begin the application of the process in the slowest quarter of the year to allow airlines to dispense with the onerous requirements for revenue guarantees, if any, in order to free up promotional funds to announce the change to the key target markets. Third, put the changes into effect and examine benefit flows with a robust tourism satellite account.

Some of the on-island sources of revenue are as follows:

- Residents pay local applicable taxes on the base price of airline tickets purchased locally.
- Accommodation taxes are increased by the amount required to cover the removal of ticket taxes.
- Increased taxes are collected from incremental visitors.
- Increased taxes are collected from increased employment.

The first tourism-dependent island economy to implement this approach to stimulating its business will not only reap the rewards of such a stimulus, but will surely create a domino effect in their region because other destinations cannot afford not to follow suit. Once the benefits begin to flow they would all do well to combine their resources and lobby for the reciprocal removal of departure taxes from the tickets of travellers from their source countries.

Today, the tourism-dependent islands of the Caribbean are operating with approximately 60% of their available accommodation occupied annually. If the stimulus provided by much lower airfares result in occupancies of 90%, the tourism contribution to GDP of these tourism-dependent economies will have risen by 50%. The effect of such an outcome on employment, and debt reduction *without any further foreign direct investment* would be unprecedented.

It is a firm conclusion that of all the initiatives available to tourism-dependent islands the one that has the most positive and powerful effect is low-priced air tickets.

References

Campaign for Better Transport (2013) *Door to Turnstile, Improving Travel Choices for Football Fans.* Campaign for Better Transport, London.

Caribbean Tourism Organization (2015) Bahamas Prime Minister Perry Christie on Caribbean Tourism. Caribbean Tourism Organization, Barbados.

European Union (2017) About the EU Size and Population. European Union.

Holder, J.S. (2013) *Caribbean Tourism*. Canoe Press, Kingston, Jamaica.

Jordan, L. (2006) Staging the Cricket World Cup 2007 in the Caribbean: issues and challenges for small island developing states. In: Robertson, M. (ed.) *Sporting Events and Event Tourism: Impacts, Plans and Opportunities*. Leisure Studies Association, Eastbourne, UK, pp. 17–42.

Jordan, L., Tyson, B., Hayle, C. and Truly, D. (eds) (2016) *Sports Event Management, The Caribbean Experience*. Routledge, Oxford, UK.

McLeod, M., Lewis, E. and Spencer, A. (2017) Re-inventing, revolutionizing and transforming Caribbean tourism: multi-country tourism institutions and a research agenda. *Journal of Destination Marketing & Management* 6(2017), 1–4.

MasterCard Advisors (2006) U.S. travelers dream about seeing Australia/New Zealand but Florida is most likely vacation destination. MasterCard Advisors, LLC. Available at: www.mastercard.com/assetfile/nc/1003/MCA/MCA/142/AssetViewServlet13.pdf (accessed 30 October 2017).

PricewaterhouseCoopers (2013) The economic impact of Air Passenger Duty. PricewaterhouseCoopers, UK.

World Bank (2015) St. Kitts and Nevis Population Total. The World Bank Group. Washington, DC. Available at: http://bpp.worldbank.org/data/exploreeconomies/st-kitts-and-nevis/2017 (accessed 12 October 2017).

3 An Altruistic Behavioural Change Approach to Sustainable Tourism when Funds are Scarce: A Narrative of a Small Business Model

Sheree-Ann Adams*

Tourism Services Ltd., St. George's, Grenada

3.1 Introduction

Since the 1983 United Nations World Commission on Environment and Development's tabling of The Brundtland Report, Our Common Future (WCED, 1987), a new business model of sustainable tourism development has been encouraged. It is one that advocates the growth of economies based on policies and practices that do not harm, and can even enhance, the environment. In spite of the many consultations held in forums such as the annual United Nations World Tourism Organization conferences, involving global stakeholders at all levels, the strategic approach to date has generally been a top-down strategic approach within B2B operations including cruise companies. This approach to change in current business operations has been led primarily by companies and organizations within the generating markets that have sought to apply and encourage the supply chain in the developed and developing countries to voluntarily change business practices where necessary towards more eco-friendly and ethical business practices. This often comes at a financial cost to grass-root suppliers that can ill-afford this expense in a seasonal cruise tourism climate. Yet the emphasis continues to be placed on 'certification/labels' to gain a competitive advantage in the marketplace and on implementing new technologies at extensive initial financial costs that are expected to produce cost savings in the long-term once the initial expense is covered. This 'lack of funds' barrier acts as a double-edged sword and often small companies in developing nations such as the Small Island Developing States (SIDS) within the Caribbean are left in a conundrum as to the trade-offs and sacrifices that need to be made to achieve the sustainability accreditation. These are conducted through annual audits from internationally

*E-mail: grenscot@hotmail.com

 Tourism Management in Warm-water Island Destinations (eds M. McLeod and R. Croes)

recognized organizations, in order to maintain preferential status for contracts with large corporations (including cruise companies) within the generating markets. This chapter presents an insight into the process and the approach currently being undertaken by a small business enterprise within the Caribbean island of Grenada.

Grenada is located in the southern Caribbean. The Caribbean archipelago is part of the world's SIDS biodiversity hotspots. The Caribbean's biodiversity remains at risk from a variety of human-induced pressures; this makes the conservation and sustainable use of biodiversity of vital importance for the regions' sustainable development. Grenada is the largest island within the tri-island state of Grenada, Carriacou and Petit Martinique. Grenada is 134 square miles and has a population of 106,000 people (UNWTO SIDS, 2013). It depends on tourism as a primary source of employment and GDP contributor. Cruise tourism is seasonal, operating from the months of October to April. One of the excursion providers to cruise ships is Tourism Services Limited (TSL). TSL is a multi-functional tourism company that started business operations in 1997 trading with the first of its adventure tour brands – the 'Adventure Jeep Tour'. Over the next ten years the company expanded its product capacity and offerings to include other activities – mountain biking (Trailblazers mountain bikes) and river tubing (Adventure River Tubing). Falling in the category of micro, small and medium-sized enterprise (MSME), the company has grown from strength to strength as one of the island's primary soft-adventure tour/excursion providers to cruise ship companies. This chapter examines the age-old topic of awarding certification or 'labels' by first looking at the issue of communication of sustainability policies and practices to consumers by companies; the use of labels in the context of the purpose and expected benefits to companies are discussed and also highlighted are the barriers experienced through the eyes of TSL. The chapter briefly outlines the alternative route currently undertaken by TSL in anticipation of achieving certification in the future.

3.2 Why Companies Communicate Sustainability Policies and Practices to Consumers

From a basic marketing perspective the success of an organization is dependent on the relationship between the product, where it is placed, how it is promoted and at what price it is available (Perreault Jr. and McCarthy, 1995). In essence there must be the right marketing mix given the wider marketing environment of political, legal, regulatory, economic/competitive, technological and societal/environmental forces (Kotler, Roberto and Lee, 2002). Therefore, by proper product/brand positioning and using strategies that target specific markets, companies can attain a competitive edge or differential advantage over other similar or substitute products in the race towards the target and ultimate goal of achieving consumer satisfaction all of the time from the purchase of a product and for a specific purpose by the consumer.

The 'four Ps' of marketing (see Table 3.1) provide the company with a balanced approach in the wider context of mass marketing to achieve consumer satisfaction given the wider marketing environment. Companies use social and environmental responsibility as a tool of brand image to position the product/company within the wider global marketplace. However, since today's market is larger given the global nature of business and consumer environments, some companies including cruise

Table 3.1. The 4 Ps and 4 Cs marketing mix approaches. (Adapted from Perreault Jr. & McCarthy, 1995 and Lauterborn, 1990.)

4 Ps	4 Cs	Approach focus
Product	Consumer needs and wants	Product: A company sells a tangible good or intangible service to satisfy what a consumer demands. Consumer: A company will only sell what the consumer specifically wants to buy.
Price	Cost	Price: The amount a consumer pays for the product and it also determines company profit and survival. Cost: Price is taken to be only part of the total cost to satisfy a want or need of the consumer.
Promotion	Communication	Promotion: All the different methods of manipulation that the seller may use to provide information to different buyers about the product. Communication: Cooperation between the seller and buyer and originates from the buyer based on needs and lifestyle.
Place	Convenience	Place: Providing the product at a place that is easily accessible by the consumer. Convenience: Based on how the target market prefers to buy.

lines take a more consumer-orientated approach to marketing in an attempt to better fit the product from a one product fits all mass marketing approach to one of niche marketing. In essence taking a consumer-orientated 'four Cs' approach (Lauterborn, 1990) (see Table 3.1) as opposed to a 'four Ps' approach towards achieving consumer satisfaction. Table 3.1 has further details on these marketing mix approaches. The strategy focus changes primarily from a manipulative promotional approach of companies as sellers to one of cooperation with consumers as the buyers through communication. This approach ensures that consumers' wants and needs are met, at an affordable cost and products are conveniently available given the era of Internet, catalogues, credit cards and phones.

The methods used by companies to communicate their sustainability policies and practices to consumers will vary depending on whether the objective is to gain a competitive advantage as an industry leader or as legitimization when the company is low performing from an ethics and/or sustainability perspective (Johnson, Scholes and Whittington, 2008). Common environmental focused strategies include 'Green marketing'; 'Green communication' and 'Green branding' (Peattie and Charter, 2003; Grant, 2007). Socially focused strategies include sponsorship, 'Cause-related marketing', 'Cause promotion', 'Corporate social marketing', philanthropy and employee volunteering (Li, McCabe and Van Der Eijk, 2011; Gatej Bradu, 2012). Companies use their sustainable development actions to show company environmental integrity and social equity principles (R. Harrison, Newholm and Shaw, 2005; First and Khetriwal, 2010; J.S. Harrison, Bosse and Phillips, 2010) so that consumers as stakeholders can identify with the companies. These strategies for communicating sustainability policies and practices are based on the thinking that consumers will associate themselves with the identity of brands/companies that are reflective of their self-identity as consumers (Lii, 2011; Gatej Bradu, 2012). Combined, these social and environmental strategies used by companies reflect the sustainability policies and practices of companies. The merits of these competitive advantage and/or legitimization communication strategies are briefly explored.

Positive consumer perceptions about the quality and nature of companies' products, their environmental awareness, and government and community relations have increasingly developed into the bases of this competition (Prahalad and Hamel, 1994; Waddock and Graves, 1997). The argument is that good company behaviour can equate to a good reputation that will in turn pay-off as increased purchases from consumers, as consumers' confidence in the product increases. However Bronn and Vrioni (2001) caution that there are studies challenging this outlook that companies can gain a competitive advantage by using marketing strategies such as 'Cause-related marketing'. For example, studies refer to increased scepticism on the part of some consumers towards companies who make claims in their company's responsible marketing efforts (Mohr et al., 2001). Others indicate consumer disbelief or mistrust due to actions that are seen to be ethically questionable (Doane, 2005). Strategies based on 'Green marketing' may be interpreted as a public relations exercise to cover-up or greenwash a company's image (Fan, 2005; Honey, 2008) or an attempt to justify a product as being socially acceptable (Doane, 2005). In essence, good corporate reputation is no guarantee of a competitive advantage as not all consumers find corporate behaviour to be of significant importance in their purchasing decisions (Boulstridge and Carrigan, 2000). Nor is good corporate responsibility a guarantee of good reputation or impacting positively on company brand image (Bhattacharya and Sen, 2004; Castaldo *et al.*, 2009). This means that for companies to increase their chances of gaining a competitive advantage by using sustainability policies and practices as attributes, companies should first be clear as to why they have made this choice in behavioural change.

It is therefore clear that there are multiple reasons as to why companies change behaviour, but the most evident can be surmised as engagement in sustainable development strategies particularly to gain a competitive advantage over existing products and to legitimize their existing products; or for altruistic reasons because they feel that it is indeed the right thing to do for 'people and planet' (El Dief and Font, 2010; Font *et al.*, 2012). Whatever the reasons for engagement, it ultimately impacts on the image of the companies and their brands from the perspective of the consumers. The Brundtland Report suggested that this change in behaviour by businesses should be for altruistic reasons – for 'people and planet', not only for the benefit of present populations, but also for the benefit of future generations, as mankind is the custodian of planet Earth. However, as time has progressed, the bottom line of the company's balance sheet has taken precedent and the justifications emphasize that today's consumers are more eco and ethically conscious and are expecting this change in business practices from all industries globally (Miller *et al.*, 2010). Also, with this behavioural change comes long-term financial savings in day-to-day operations, but the greatest emphasis has been placed on the competitive advantage a company will have over similar companies if they are the first (or among the first) within their immediate business community to receive 'green' credentials (labels). Many companies follow the latter path at great financial sacrifice, some take to greenwashing and social washing, while others merely defer a 'start date' to a period when the tourism destination authorities deem 'green credentials' to be compulsory.

3.3 Purpose and Execution of Labels

The process of validation and the resulting certification are effective ways for visitors to have assurance as to which businesses are acting sustainably. While there is continuous debate as to whether this is actually true or not (Dodds and Joppe, 2005; Font and Epler Wood, 2007; Graci and Dodds, 2015; Responsible Travel, 2017), what is evident is that regardless of size, it has impacted on business operations globally and the process is in a state of continuous redevelopment. To quote Xavier Font, formerly of Leeds Beckett University International Center for Responsible Tourism, 'The process of validation has improved the practices and transparency of certification programs that passed and those that didn't, by improving their organizational systems, reviewing criteria, improving verification practices and membership benefits' (Stanford, 2014). These redevelopments and process improvements are necessary due to the underlying purpose of the labels. That is to generally serve as a benchmarking system for businesses that use a set of standards to guide sustainability improvements within the business and destination by extension. The sustainability progress is measured through a procedure that evaluates the performance of a business against a set of agreed sustainable tourism practices. In essence, the 'label' is proof that the business has achieved the required level of sustainability practices. However, it is important to note that labels are generally voluntary. They are expected to be user-friendly and designed to inspire more efficient and sustainable ways of doing business.

Which label or certification programme is selected is generally an individual business choice or may be influenced by the island's governing tourism authority. Some destinations choose to use external existing schemes. For example, Visit England uses three primary programmes: (i) Green Tourism Business Scheme (throughout the UK); (ii) ISO20121 Sustaninable Event Management Certification, the International Standard for sustainable events awarded by SGS; and (iii) Environmental Quality Mark (in use in the Peak District and Staffordshire) (Stanford, 2014). Other countries such as the Seychelles (Seychelles Tourist Board, UNDP and GEF, 2012) and Costa Rica (Pearson, 2006) have chosen to implement their own standards and labels that they believe meet international standards and are reflective of the positive ethical behaviour and environmental stewardship of their stakeholders. Green Globe, Travelife and Blue Flag are best known from an international tourism industry certification programme (Graci and Dodds, 2015). In the Caribbean, the two most prominent accreditation companies are Green Globe, which is endorsed by the Caribbean Hotel and Tourism Association, and Sustainable Travel International (STI).

In September 2014, STI, along with Royal Caribbean Cruise Lines (RCCL), the Caribbean Tourism Organization (CTO) and the Organization of American States (OAS) formed the Caribbean Sustainable Tourism Alliance with the primary objective of assessing and assisting the Caribbean destinations with the development and implementation of their Sustainable Tourism Labels (STL). STI has also teamed up with RCCL to assess and certify the land and marine tour operators within the region in their efforts toward the greening of the cruise tourism product supply chain, of which STL is part. Most recently, Royal Caribbean in keeping with its 'continuous improvement' mantra has teamed-up with World Wildlife Fund in a 5-year partnership to protect the oceans (RCL CORPORATE, 2016, 2017(b)). The target is for 1,000 of its destination tours to come from operators certified as environmentally, socially and culturally sustainable practitioners by the end of 2020 (RCL CORPORATE, 2017(a)). Royal Caribbean is arguably the most proactive of the large global cruise ship brands seeking to achieve a sustainability 'seal of approval' from the shore excursions supply chain. This is one of the company's many sustainable cruise tourism development initiatives and their commitment to implementing strategies to achieve 'green' and 'ethical' cruise ship brands is commendable and noteworthy.

However, although at this time Grenada does not have its own official STL that is recognized internationally, endorsed by a third-party auditor and endorsed and/or enforced by the Grenada tourism authority and/or the ministry of environment, there are individual tourism businesses that have sought accreditation from recognized certifying companies such as Green Globe. As of January 2016, there are five Green Globe Accredited Members (Green Globe, 2016) of a total of 88 accommodation properties in the tri-island state (GTA, 2016). All five hotels are members of the Grenada Hotel and Tourism Association. As of December 2016, none of the cruise excursion suppliers on the island have achieved the RCCL recommended STI STEP certification for shore excursion providers. This low uptake in sustainability certification is primarily due to the high cost associated with enrolling and annual audits to attain the desired 'label'.

Running an audit programme for a label, as with any business operation, is a costly exercise as it requires a multitude of resources that include funding, audit techniques, auditor training, auditor development, able auditors, supporting expertise, the audit programme; and general support resources such as travelling time, transport and accommodation (Wealleans, 2005). These costs are generally passed on to the client, resulting in the high cost often associated with engaging auditors to achieve certification. While the cost will vary from one verifying body to the next, this high cost associated with the certification has generally made it accessible notably to larger hotels and tourism businesses (Honey, 2002). For example, it would cost a medium-sized hotel approximately \$20,000–\$40,000 for its certification inclusive of travel, staff training and consultation (Toth, 2002). The STI Step Shore Excursions certification starts with an attractive \$750 website enrolment fee, but can cost an annual minimum of \$5,000 for auditors' fees, plus travel and accommodation expenses in addition to costs of staff training, consultations and the use of templates. At these prices and country currency exchange rates, certification is more often than not placed out of the financial reach of many SME's that make up a significant proportion of the largely seasonal revenue-generating cruise tourism business within the Caribbean's tourism industry. It is with good reason that countries such as Costa Rica have offered the certification/label gratis for its stakeholders, to ensure that all its stakeholders adequately represent the destination's standards; and to allow for recourse in the event that standards are breeched. This does not mean that grass-root shore excursion suppliers such as TSL have not entertained the thought or started the process on their own.

By having a better understanding of what the certification process actually entails, companies have determined the actual initial costs, return on the investment and alternative options to start the process to make the company 'verification ready' in an effort to reduce the costs associated with the services of the labelling organizations. The steps in all labels generally follow a similar five step format for implementation and evaluation (see Table 3.2). These are as follows:

- Step 1. Communication.
- Step 2. Application.
- Step 3. Assessment.
- Step 4. Verification.
- Step 5. Award/Certification (Font, 2002).

What may differ are the grading, point system and verification fees associated with the individual labels. RCCL has indicated that in addition to the preferred STI STEP programme, the company would also accept the certification of other labels to avoid companies going through the expense of multiple labels. These additional labels are: Certification in Sustainable Tourism (CST) from Costa Rica based tour operators and excursions providers; EarthCheck from South Pacific and Asia based large tour operators and small excursion providers; Rainforest Alliance from small-scale operations in Latin America; and Travellife from European-based tour operators that have an extensive line of sub-contracted excursions in addition to the internationally recognized Green Globe, which certifies not only hotels, but also large tour operators and small excursion providers as well as the longer standing International Standards Organization (ISO) 14001 (STI, 2012). Arguably, ISO 14001 series is the most popular of the environmental management standards, because of the advantage of its versatility and ability to be applied in many industries including tourism and within various industry sectors (Honey, 2002). This strength is also criticized as a weakness as this approach is too broad to provide an accurate measurement of the environmental impacts within different industries. RCCL recognizes that this standard is more relevant to large tour operators and does not adequately cover sub-contracted excursions. RCCL's initial target was to have 50% of all the shore excursions available

Table 3.2. Moving towards certification: altering the conventional process. (Adapted from Font, 2002, p. 202.)

Conventional steps of the label (certification) process	TSL chosen pathway towards achieving certification
Step 1: Communication	Step 1: Acknowledge issue; communication
Step 2: Application	Step 2: Self-assessment; preparation of a sustainable tourism plan; implementation of the sustainable tourism plan
Step 3: Assessment	Step 3: Select a label and apply for verification
Step 4: Verification	Step 4: Verification
Step 5: Award/Certify	Step 5: Award/Certify
Repetition of Steps 3, 4 and 5 on an annual basis to maintain accreditation	Repetition of Steps 3, 4 and 5 on an annual basis to maintain accreditation

to their passengers third-party verified to an internationally recognized sustainability standard by the end of 2015 as part of the company's destination stewardship programme (RCCL, 2010; STI, 2010). No target date was given for the other 50% to comply, but they had incentivised participation by providing certified operators with exclusive marketing benefits through the provision of a Sustainable Shore Excursions icon on the pre-departure and on-board marketing channels (RCCL, 2010) implying that these were the excursions of preference. The RCL-WWF 5-year partnership targeting 1,000 sustainably certified tours by 2020 appears to be the continuation of the initial RCCL–STI partnership strategy. As of June 2017, none of the RCCL Grenada Shore Excursions have this icon displayed on the RCCL website, suggesting that the Grenada suppliers have yet to voluntarily comply.

This emphasizes the importance of the first step of the accreditation process – the acknowledgement of the issue and communication stage. This stage is the most important as clear communication is vital to get stakeholders to buy in to the concept of certification through the label message and understanding the benefits of the certification process. Users have to attain a level of confidence in the scheme and general agreement as to the value of the process of validation towards improving the company's systems. This generally involves various levels of communication, stakeholder engagement and awareness-raising. For example, RCCL and STI have presented on their Destination Stewardship and Sustainable Shore Excursions programmes at conferences and trade show panels to engage the destination tourism authorities, tour operators and excursion suppliers. For any label, the indicator of successful communication is businesses signing up for assessment or technical assistance.

For grass-root shore excursion companies, as with any tourism business, a key element in the final decision to undertake the time and cost involved in the certification process, more often than not, is the potential return on investment (ROI) to the company. To convince enterprises of the benefits of any certification or labeling scheme, ROI must be demonstrated. There are two main benefits which are generally communicated to tourism enterprises: marketing and cost savings.

Marketing

The key premise is that certification is expected to provide operators with a competitive advantage in the market place over uncertified enterprises. For example, with Destination Certifications such as the Seychelles, certified properties are made the promise of being highlighted or placed in areas of prominence and listed on websites, brochures and other marketing material and given precedence in promotion at trade fairs (Seychelles Tourist Board *et al.*, 2012). RCCL's offer is limited to the company's pre-departure and on-board marketing channels (RCCL, 2010). The underlying justification being that the label is also expected to provide increased guest satisfaction levels with the communication of sustainability achievements of the organization, as consumer product confidence will be increased through label endorsement as a 'good' as opposed to a 'bad' product/operation. TSL acknowledges that 2:1 cruise consumers have a preference for companies with sustainability policies and practices, but when placed in the context of attribute importance, sustainability policy and practices ranks fifth after product price, product quality, cruise itinerary/destinations and cruise duration (Adams and Font, 2017; Adams *et al.*, 2017). This suggests that a label does not guarantee cruise customers would use a label as the primary product choice determinant, but may use the label as the influencing factor when choosing between multiple similar products with no price and quality differentiation. Therefore, TSL's decision to initiate steps towards sustainability certification leans towards altruistic reasons as opposed to anticipated competitive advantage in the arena of shore excursions.

Cost savings

Cost savings is the second benefit that is much emphasized and anticipated. The expectations are that company utility costs are likely to decrease due to increased awareness, monitoring and use of improved technology. Staff replacement costs are likely to decrease as research shows sustainability initiatives build staff motivation and increase retention rates (Weber, 2008). These proclamations are made based on the assumption that the company signing on to the label does not already engage in any of the practices that are considered to be sustainable for the business to operate eco-efficiently and ethically in the long term. Furthermore, the only cost saving to any company already engaging in eco-efficiency and ethical practices (sustainability practices) would be a shorter auditing process (reduced billing for auditors' time and accommodation expenses) to achieve the certification by the desired label. Companies are audited by the certifying body on the request of a company (application is made to the certifying body) and the company is assessed once the statement of compliance with the labels 'must criteria' is signed and the company has performed the desired sustainability self-assessment. The self-assessment form is generally the same form used by assessors, only instead of points the organization's manager notes 'yes' or 'no' for each criterion based on whether they want to be assessed on these criteria. The certification body will review the self-assessment form to ensure the organization has selected sufficient criteria to be able to pass the assessment. To maximize the chances of success, it is recommended that organizations should aim to be assessed on 25% more points than they actually need to pass (Seychelles Tourist Board *et al.*, 2012). The certifying body in essence facilitates a process, but the onus remains on the client as to how much benefit will actually be gained from the process and recognition of the label by current and future customers.

Furthermore, in assessing the ROI and benefits, companies must bear in mind that certification is not indefinite. Organizations are required to be reassessed generally within a 12- to 18-month period following the last assessment and date of the award. Generally, if more than two years have passed and the organization does not re-apply to be assessed the certification is deemed lapsed. The company will no longer be eligible to use the logo in its marketing and communications, including promotional material produced by its B2B clients. Company names and logos are also removed from the awarding bodies' websites, a benefit of membership. Companies must therefore ensure that the cost of certification is included in its annual expense projections and budgeted for so as to ensure continuity and membership benefits. Labels should therefore be treated as a long-term investment and not entered into without a proper investigation of the process and the actual financial and projected expense to the company.

3.4 Being Penny Wise – Using Existing Free Assessment and Guidelines Tools to Start the Sustainability Process as Part of a Strategic Plan Towards Certification

Many small companies are often so consumed with day-to-day financial survival and fail to look for an alternative solution to enable social and environmental changes in their operational business practices. The environmental issues such as littering, reef damage and the eating of turtle eggs, turtle meat and monkey meat (wild meat) are more often than not a result of a lack of education and awareness of persons not employed directly in tourism. Few changes are required directly by grass-root tourism companies as they are held accountable and governed by local employment regulations and any existing environmental laws that protect the country's natural assets including flora and fauna. More often than not, grass-root companies have taken ethical and eco-efficiency measures primarily to raise awareness of the value the natural environment to the sustainability of tourism. The focus is on the protection of the environment and the natural resources, to ensure that the employees from the communities in which the businesses operate clearly understand the significance of their natural working, and often home, environment to the bottom line and survival of the business.

The challenge is the process of accumulation of the required data for certification as data should be collected over extended periods of time and organized in a manner that can be presented on request to customers, whether in B2C or B2B business scenarios. The guideline for the presentation of the data is therefore crucial. The most effective starting point is to have a strategic plan as to the methodology to be engaged. Once there is an understanding of the objectives of a sustainability audit, making the criteria the central focus of a sustainable tourism strategic plan for the company is the next important step. If the objective is to 'get the company in order' to enable the audit process for certification by a third-party verification organization, in minimum time possible and by extension at the lowest cost possible, then the following steps are within any company's reach regardless of size.

Conduct a self-assessment

The self-assessment is traditionally the second step within the application for the certification/label process. This process provides the auditor with insight as to the company's stage in the sustainability process. Third-party verification organizations would not provide this document unless the prospective client has made an initial payment and is committed to the programme. It is, however, worth noting that the self-assessment questions are in line with the benchmark criteria used in the audit. Self-assessment questionnaires are available online. One example is that of the Vancouver Island University Sustainable Tourism Self-Assessment Tool, available online in PDF form at the university's website.[1] It is worth doing the self-assessment exercise to establish the 'sustainability position' of the company. These self-assessments are subjective but give an idea of the areas the auditors will most likely be focusing on during the audit in the certification process (see Fig. 3.1).

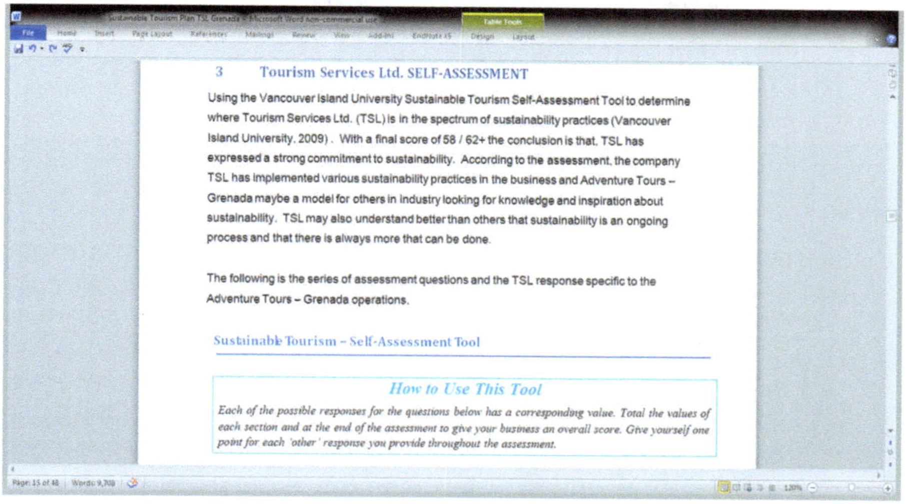

3 Tourism Services Ltd. SELF-ASSESSMENT

Using the Vancouver Island University Sustainable Tourism Self-Assessment Tool to determine where Tourism Services Ltd. (TSL) is in the spectrum of sustainability practices (Vancouver Island University, 2009). With a final score of 58 / 62+ the conclusion is that, TSL has expressed a strong commitment to sustainability. According to the assessment, the company TSL has implemented various sustainability practices in the business and Adventure Tours – Grenada maybe a model for others in industry looking for knowledge and inspiration about sustainability. TSL may also understand better than others that sustainability is an ongoing process and that there is always more that can be done.

The following is the series of assessment questions and the TSL response specific to the Adventure Tours – Grenada operations.

Sustainable Tourism – Self-Assessment Tool

How to Use This Tool

Each of the possible responses for the questions below has a corresponding value. Total the values of each section and at the end of the assessment to give your business an overall score. Give yourself one point for each 'other' response you provide throughout the assessment.

Fig. 3.1. Snap-shot of TSL self-assessment result summary.

Set out the strategic sustainability plan using the Global Sustainable Tourism Council Criteria

Today's sustainability certification process is once again under the auspices of the United Nations. One of the more recent developments has been the recognition and validation by the United Nations and the Global Sustainable Tourism Council (GSTC), of labels that are providing a programme that uses and conforms to the GSTC standards and criteria. This is a great improvement to the situation of 12 years ago when 70 similar programmes were identified, all with a lack of comparable standards and criteria (WTO, 2002). This should reduce much of the market confusion surrounding which labels are best or appropriate, but to date has had little impact on the high start-up costs associated with certification process for companies seeking accreditation. These criteria are readily available for free online, at the GSTC website[2] and should form the basis of the benchmark criteria in a company's sustainability audit plan. Therefore, companies can start the process of evidence identification (i.e. cross reference notations of the location of the evidence in existence) and action shortfalls using the GSTC criteria, indicators and guidance notes knowing with confidence that the label ultimately engaged should be using the same guidelines and criteria.

Set-up spread sheets to capture data and monitor progress

Capturing and monitoring of the data are by far the most time consuming and laborious of the processes. While most companies will be in possession of recorded data, its format will differ depending on its purpose, established by the specific objectives of the initial data capture process. Small companies such as TSL currently use spreadsheets for accounting and invoicing data as the focal point of any strategic planning. Given the highly seasonal nature of the business, focus is on breaking even annually as off-season operating costs are supplemented by in-season profits. Highest priority is given to the payment of employee wages, repayment of commercial bank loans often at 12% interest and the maintenance of tour-related equipment as safety supersedes any other form of marketing or good will when it comes to company/brand image.

In addition, there are social acts of kindness that are not currently formally recorded in the captured data and quantified with an associated financial cost to the company. These acts of kindness now have to be formalized, recorded, quantified and labelled as contributions of social responsibility, removing the often impromptu heartfelt gestures that have strengthened the relationship between the company and the communities in which it operates. Larger organizations, including hotels/individuals in Grenada and the wider Caribbean, engage in these corporate giving and philanthropic acts formally, generally receiving tax rebates in lieu of funds invested in social programmes. Small grass-root companies in developing countries extend this practice of social giving on a daily basis but the extent of these mostly impromptu 'donations' are rarely recorded. These acts may not always be monetary but an act of goodwill in gesture, such as carrying some local farmers' produce to market as the tour vehicle returns to base in the capital; donating the use of the tour vehicles in motorcades to raise funds for persons in need of medical attention abroad; or collecting used school books from friends and family to pass on to other children within the rural communities in which the excursions operate. Unfortunately, what was once the basic principle of humanity – to treat others as one would wish to be treated – is overshadowed by the requirement to jump up and down and shout, 'Look at us, look at the good we have done and are doing!' Social media and labels are used as the forerunners in marketing platforms for this 21st century business practice.

Staff cooperation is essential

Make sure all staff are on board and appoint a sustainability point person if there isn't already one. The person responsible for product quality control is the logical existing position. This will mean additional responsibilities and will more often than not require additional compensation in wages. However, the biggest challenge can often be getting the employees living and working in the natural environments, often themselves living in the poorer rural communities, to change the day-to-day way of living. For example, recycling of plastics is a normal activity in the form of re-use as opposed to recycling in a purpose-built facility. A simple example such as this, suggests that some important sustainability issues may not be in line with the benchmark mark criteria used by the auditing company. It is important for the company to continue with practices that are ethical and eco-efficient to each individual business, but to bear in mind that these may not be taken into consideration and extra points will not be given for those initiatives. The behavioural change has to be a team effort and should put the indigenous population (nationals) and the environment as the central focus of the outcome impacts.

Utilize your sustainability strategic plan beyond the label purposes to improve your ROI

Use the sustainability strategic plan to approach the local government for concessions/tax holidays on new eco-efficiency products and to improve the business's social

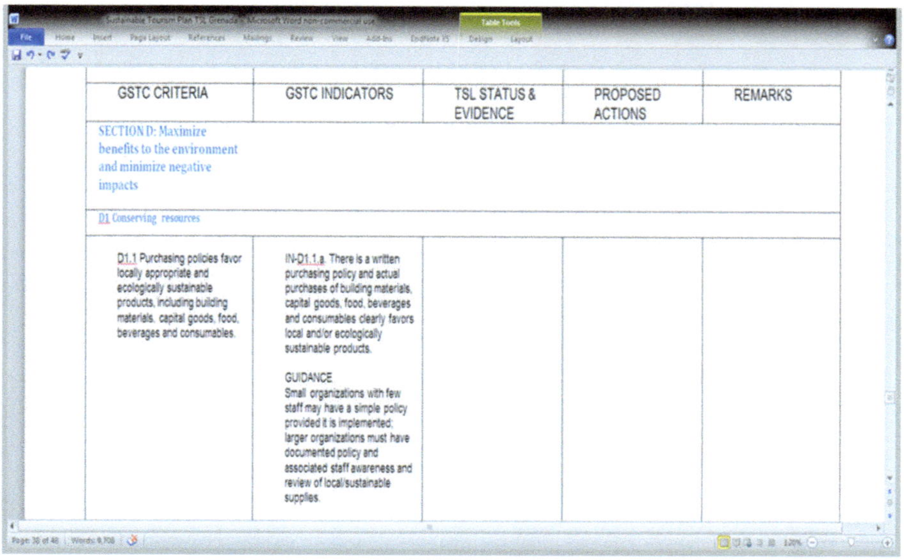

GSTC CRITERIA	GSTC INDICATORS	TSL STATUS & EVIDENCE	PROPOSED ACTIONS	REMARKS
SECTION D: Maximize benefits to the environment and minimize negative impacts				
D1 Conserving resources				
D1.1 Purchasing policies favor locally appropriate and ecologically sustainable products, including building materials, capital goods, food, beverages and consumables.	IN-D1.1.a. There is a written purchasing policy and actual purchases of building materials, capital goods, food, beverages and consumables clearly favors local and/or ecologically sustainable products. GUIDANCE Small organizations with few staff may have a simple policy provided it is implemented; larger organizations must have documented policy and associated staff awareness and review of local/sustainable supplies.			

Fig. 3.2. Snap-shot of TSL sustainability action plan format using GSTC criteria as guidelines.

status of giving on island (see Fig. 3.2) by formalizing any existing acts of giving. For example, 'adopt' a farmer or school as part of a community support project. Keep accurate accounts of all money injected into social projects such as these, as they are often tax deductible. The former, eco-efficiency has more direct benefit on the grass-root company's bottom line if import taxes are waived as the cost of eco-efficient equipment is often very costly and subject to high importation costs and taxes. For example, the use of catalytic converters that reduce emissions from older vehicles are an alternative for developing nations such as Grenada where low-carbon alternative fuels are not yet available nor expected to be in the foreseeable future. Many companies in developing countries cannot afford the luxury of changing vehicles every 3 years due to the high interest rates on borrowing money from commercial banks and the high vehicle importation duty that is often between 80–130% of vehicle cost inclusive of shipping. It is imperative to explore all options to maximize the benefits from the investment in sustainability certification and labels.

3.5 Conclusion

Sustainable tourism development strategies continue to be an evolving topic. Current evidence suggests that the most notable changes towards ethical and eco-friendly behaviour are within the day-to-day management and operation policies and strategies of corporations including cruise companies such as RCCL. Transparency and accountability are therefore the primary policing mechanisms to ensure that sustainability policies and practices that are used by corporations are not only far reaching to include their supply chain, but also conducted in a manner that is in the best interest of 'people and planet'. This is the underlying principle of the Brundtland Report. Yet the documented evidence and proposed strategies clearly suggest that the emphasis is on marketing and cost savings that can positively impact on a company's brand image and ROI. In essence, the message conveyed is to take the sustainability route as it is good for the company's bottom line. Labels (certification) have become a key marketing tool in the global sustainability strategy drive. It serves primarily as an indicator to consumers and business partners globally, supposedly of a company's commitment to ethical and eco-friendly behaviour in an age where technology has allowed companies few places to hide and provides platforms for companies to engage current and potential customers through the use of sustainability marketing tools.

Small, owner-operated companies are expected to operate on the same platforms and utilize the exact strategies as large multinational corporations. These corporations originated and are housed in the developed nations but conduct business within the wider global economy. Nothing has really changed in the chain of command; the approach is still top down (developed nations setting tourism policies and strategies) in spite of the many exercises of stakeholder consultations that have included representatives from the developing nations. Developing nations, and by extension the local tourism businesses, remain financially dependent on the developed nations. The developing nations and businesses within continue to make all efforts to comply with the strategies (both compulsory or voluntary) that have been developed by experts, many of whom were not born into poverty, and who have never owned nor operated a small business in a developing country.

The onus therefore remains on the small tourism businesses such as TSL to find ways of 'keeping up with the Joneses' with regards expectations; in spite of any additional financial burdens, strategies such as sustainability labels and the cost third-party audits may present in a highly seasonal business environment, where priorities differ and the cost of borrowing, gives small companies a slim chance of even reaching a 15% ROI that corporations can often comfortably take for granted. This chapter has provided readers, themselves small business owners with some useful website links and a suggested strategy to start the process and gain a better understanding of what is generally involved in the process of sustainable tourism certification. These recommendations do not solve the underlying issue of funding certification, but should allow for better planning towards the inevitable, which would see all businesses certified or accredited under a scheme.

Notes

[1] The Vancouver Island University Sustainable Tourism Self-Assessment Tool can be accessed in PDF form at: http://web.viu.ca/sustainabletourism/Self%20Assessment%20Tool%20June%204.pdf (Vancouver Island University, 2009)

[2] The sustainability certification criteria can be accessed free of charge at: http://www.gstcouncil.org/images/pdf/HTO-CRITERIA_and_INDICATORS_6-9-14.pdf (GSTC, 2012)

References

Adams, S. and Font, X. (2017) Purchasing attributes for cruise passengers. In: Dowling, R.K. and Weeden, C. (eds) *Cruise Ship Tourism*, 2nd edn. CAB International, Wallingford, UK, pp. 290–303.

Adams, S., Font, X. and Stanford, D. (2017) All aboard the corporate socially and environmentally responsible cruise ship: a conjoint analysis of consumer choices. *Worldwide Hospitality and Tourism Themes* 9(1).

Bhattacharya, C.B. and Sen, S. (2004) Doing better at doing good: when, why and how consumers respond to corporate social initiatives. *California Management Review* 47(1), 9–24.

Boulstridge, E. and Carrigan, M. (2000) Do consumers really care about corporate responsibility? Highlighting the attitude-behaviour gap. *Journal of Communication Management* 4(4), 335–368.

Bronn, P.S. and Vrioni, A.B. (2001) Corporate social responsibility and cause-related marketing: an overview. *International Journal of Advertising* 20, 207–222.

Castaldo, S., Perrini, F., Misani, N. and Tencati, A. (2009) The missing link between corporate social responsibility and consumer trust: the case of fair trade products. *Journal of Business Ethics* 84(1), 1–15.

Doane, D. (2005) The Myth of CSR. The problem with assuming that companies can do well while also doing good is that markets don't really work that way. *Stanford Social Innovation Review*. Stanford Graduate School of Business, California, pp. 23–29.

Dodds, R. and Joppe, M. (2005) CSR in the Tourism Industry? The Status of and Potential for Certification, Codes of Conduct and Guidelines. CSR Practice Foreign Investment Advisory Service Investment Climate Department, pp. 1–60. Available at: http://siteresources.worldbank.org/INTEXPCOMNET/Resources/CSR_in_tourism_2005.pdf (accessed 1 January 2015).

El Dief, M.E. and Font, X. (2010) The determinants of hotels' marketing managers' green marketing behaviour. *Journal of Sustainable Tourism* 18(2), 157–174.

Fan, Y. (2005) Ethical branding and corporate reputation. *Corporate Communications. An International Journal* 10(4), 341–350.

First, I. and Khetriwal, D.S. (2010) Exploring the relationship between environmental orientation and brand value: is there fire or only smoke? *Business Strategy and the Environment* 19(1), 90–103.

Font, X. (2002) Environmental certification in tourism and hospitality: progress, process and prospects. *Tourism Management* 23(3), 197–205.

Font, X. and Epler Wood, M. (2007) Sustainable tourism certification marketing and its contribution to SME market access. In: Black, R. and Crabtree, A. (eds) *Quality Assurance and Certification in Ecotourism*. CAB International, Wallingford, UK, pp. 147–163.

Font, X., Walmsley, A., Cogotti, S., McCombes, L. and Hausler, N. (2012) Corporate social responsibility: The disclosure-performance gap. *Tourism Management* 33(6), 1544–1553.

Gatej Bradu, C.P. (2012) CSR and Consumers' Expectations Towards It – A Marketing Perspective. *EuroEconomica* 31(1), 7–13.

Graci, S. and Dodds, R. (2015) Certification and labeling. In: Hall, C.M., Gossling, S. and Scott, D. (eds) *The Routledge Handbook of Tourism and Sustainability*. Routledge, London and New York, pp. 200–208.

Grant, J. (2007) *The Green Marketing Manifesto*. Wiley, Chichester, UK.

Green Globe (2016) Grenada Members. Available at: www.greenglobe.com/members/caribbean/ (accessed 22 January 2016).

GSTC (2012) Global Sustainable Tourism Council Criteria and Suggested Performance Indicators for Hotel and Tour Operators (version 2). Available at: http://www.gstcouncil.org/images/pdf/HTO-CRITERIA_and_INDICATORS_6-9-14.pdf (accessed 22 January 2016).

GTA (2016) Accommodation by Category For January 2016 (Public, Excel).

Harrison, J.S., Bosse, D.A. and Phillips, R.A. (2010) Managing for stakeholders, stakeholder utility functions, and competitive advantage. *Strategic Management Journal* 31(1), 58–74.

Harrison, R., Newholm, T. and Shaw, D. (2005) *The Ethical Consumer*. Sage Publications, London.

Honey, M. (2002) *Ecotourism & Certification: Setting Standards in Practice*. Island Press, Washington, D.C.

Honey, M. (2008) *Ecotourism and Sustainable Development: Who Owns Paradise*, 2nd edn. Island Press, Washington, D.C.

Johnson, G., Scholes, K. and Whittington, R. (2008) *Exploring Corporate Strategy*, 8th edn. Pearson Education Limited, Essex, UK.

Kotler, P., Roberto, N. and Lee, N.R. (2002) *Social Marketing: Improving the Quality of Life*. Sage, Thousand Oaks, California.

Lauterborn, B. (1990) New marketing litany: four Ps passe: C-words take over. *Advertising age* 61(41), 26.

Li, C., McCabe, S. and Van Der Eijk, C. (2011) *Understanding the Choice Heuristic of Chinese Long-haul Outbound Tourists*. Paper presented at the Consumer Behaviour in Tourism Symposium 2011

Lii (2011) The effect of corporate social responsibility (CSR) initiatives on consumers' identification with companies. *African Journal of Business Management* 5(5), 1642–1649.

Miller, G., Rathouse, K., Scarles, C., Holmes, K. and Tribe, J. (2010) Public understanding of sustainable tourism. *Annals of Tourism Research* 37(3), 627–645.

Mohr, L.A., Webb, D.J. and Harris, K.E. (2001) Do consumers expect companies to be socially responsible? The impact of corporate social responsibility on buying behaviour. *The Journal of Consumer Affairs* 35(1), 45–72.

Pearson, K. (2006). Sustainable Business Development in Costa Rica.

Peattie, K., and Charter, M. (2003) Green Marketing. In: Baker, M.J. (ed.) *The Marketing Book*, 5th edn. Butterworth-Heinemann, Oxford, UK, pp. 726–756.

Perreault Jr., W.D. and McCarthy, E.J. (1995) *Basic Marketing: A Global Managerial Approach*. Times Mirror Higher Education Group, Burr Ridge, Illinois.

Prahalad, C.K. and Hamel, G. (1994) Strategy as a field of study: Why search for a new paradigm? *Strategic Management Journal* 15(S2), 5–16.

RCCL (2010) Destination Stewardship Program Available at: http://www.destinationsfortomorrow.org/ (accessed 30 October 2015).

RCL CORPORATE (2016) Royal Caribbean and World Wildlife Fund Unite in 5-Year Partnership. 1–3. Available at: http://www.rclcorporate.com/royal-caribbean-and-world-wildlife-fund-unite-in-5-year-partnership/ (accessed 9 October 2017).

RCL CORPORATE (2017a) RCL wants Tour Operators with a Sustainable 'Seal of Approval'. (April 4, 2017), 1–2. Available at: http://www.rclcorporate.com/rcl-wants-tour-operators-with-a-sustainable-seal-of-approval/

RCL CORPORATE (2017b) A Partnership to Help Protect Our Oceans. 1–2. Available at: http://www.rclcorporate.com/a-partnership-to-help-protect-our-oceans/ (accessed 30 May 2017).

Responsible Travel (2017) Accreditation Schemes: Why I might create a sustainable tourism certificate scheme. *Accreditation Schemes*. Retrieved 09/06/2017.

Seychelles Tourist Board, UNDP and GEF (2012) Seychelles Sustainable Tourism Label Assessors' Manual *Assessors' Manual Version 3.0. June 2012*, Seychelles Tourist Board, pp. 1–32.

Stanford, D. (2014) Keep it real for destinations. 7 common mistakes to avoid when telling your destination's sustainable stories. Available at: https://www.youtube.com/watch?v=5VcGDonuNpU (accessed 1 January 2015).

STI (2010) STEP: Sustainable Shore Excursions Program. Available at: https://sustainabletravel.org/our-work/royal-caribbean-sustainable-shore-excursions/step/ (accessed 30 October 2013).

STI (2012) Royal Caribbean Cruises Ltd. Sustainable Shore Excursions Program. Available at: https://sustainabletravel.org/our-work/royal-caribbean-sustainable-shore-excursions/ (accessed 30 October 2017).

Toth, R. (2002) Exploring the concepts underlying certification. In: Honey, M. (ed.) *Ecotourism & Certification: Setting Standards in Practice*. Island Press, Washington, DC, pp. 73–102.

UNWTO SIDS (2013) Tourism in Small Island Developing States (SIDS): Building a more sustainble future for the people of islands. Available at: http://cf.cdn.unwto.org/sites/all/files/docpdf/tourisminsids.pdf (accessed 30 October 2015).

Vancouver Island University (2009) Sustainable Tourism – Self-Assessment Tool. Available at: https://web.viu.ca/sustainabletourism/Self%20Assessment%20Tool%20June%204.pdf (accessed 12 November 2014).

Waddock, S.A. and Graves, S.B. (1997) The corporate social performance-financial performance link. *Strategic Management Journal* 18(4), 303319.

Wealleans, D. (2005) *The Quality Audit for ISO 9001:2000 A Practical Guide.* Ashgate Publishing, Farnham, pp. 91–112.

Weber, M. (2008) The business case for corporate social responsibility: A company-level measurement approach for CSR. *European Management Journal* 26, 247–261.

WTO (2002) *Voluntary initiatives for sustainable tourism.* World Tourism Organization, Madrid, Spain.

WCED (1987). Report of the World Commission on Environment and Development: Our Common Future. Available at: www.un-documents.net/our-common-future.pdf: 1–300 (accessed 30 October 2017).

4 Tourist Motivation and Demand for Islands

Robertico Croes* and Jorge Ridderstaat

University of Central Florida, United States of America

4.1 Introduction

Tourism is considered an important engine for economic growth (Croes, 2011; Ridderstaat, Croes and Nijkamp, 2014; Brida, Cortes-Jimenez and Pulina, 2016). In particular, island destinations have demonstrated that tourism can propel fast and relatively high economic growth (Lanza and Pigliaru, 2000; Brau, Lanza and Pigliaru, 2007; Seetanah, 2011; Croes, 2013), overcoming size constraints. The positive relationship between tourism development and economic growth is anchored in the tourism-led-growth hypothesis (TLG), which improves efficiency through employment, industry generated income and terms of trade. In addition, tourism powers economic activities that prompt tourism to grow at rates comparable to, or faster than, other economic sectors (Seetanah, 2011). Based on this assertion, the TLG hypothesis claims that resources should be allocated to support tourism activities more so than other economic sectors.

The Caribbean region has embraced tourism, focusing on increasing their permanent and transient (tourists) population (Buswell, 2011). However, empirical results suggest that tourism benefits only seem clear in the short term, while the long-run benefit prospects are ambiguous. Long-run tourism benefits may be clouded by the so-called 'sluggish development' due to the Dutch disease[1] and the non-linearity aspects of tourism development (Croes, 2011; Pratt, 2015). Many island destinations in the Caribbean region desperately need growth, though they do not get the maximum advantage from their tourism potential because they lack empirical knowledge of the tourism market and the factors affecting the demand for tourism (Croes, 2011). Instead, they tend to rely on anecdotal and ad hoc advice in designing and implementing their tourism policy. Thus, tourism development programmes may fail because they have not evaluated these factors properly.

*E-mail: robertico.croes@ucf.edu

Thorough studies of the dimensions of demand could provide these countries with accurate market information, which would allow them to base their tourism policy decisions on accurate and scientifically sound analyses. Consequently, analyses could assist them in developing their full tourism potential, which, of course, would help strengthen their economies. Island destinations situated in the Caribbean Basin are facing daunting challenges in terms of population growth; they are in desperate need of fast and furious growth if they want to overcome the poverty threat. Such strengthening will occur because of market-based development; i.e. the market in those countries will respond naturally to the benefits afforded by the presence of additional tourist attendance. That is, tourism-related businesses (including small and micro enterprises) throughout the economy should benefit as the increased tourist spending ripples through the economy.

This chapter sheds light on the characteristics that determine the demand for island tourism, by assessing the motivational and economic aspects of tourists choosing to visit island destinations. The chapter focuses on the various demand models that are suitable for smaller sized islands, and whether specific determinants of demand (e.g. economic versus psychological) need to be addressed to support island tourism. A review of the nature of demand for the Caribbean tourism product, including an emerging tourist destination such as Cuba, will be included. In particular, the analysis will include Aruba, the Bahamas, Barbados, Jamaica, the Dominican Republic and Cuba.

4.2 Warm-water Destinations Confronting Headwinds

The Caribbean region consists of one of the largest insular densities in the world and depends on tourism more than any other region of the world. Tourism is the single largest earner of 16 of the 29 countries and sun, sand and sea are still an integral part of the region's charm and resource with which to attract tourists. One in eight jobs in the Caribbean is currently supported by tourism and by 2025 this is expected to increase to one in every seven jobs – providing for nearly 2.8 million jobs. According to the World Travel and Tourism Council (2015), tourism economic activity in the Caribbean by 2025 should expand to nearly $74 billion and generate tourism receipts in excess of $25 billion. These statistics clearly articulate the relevance of tourism for the Caribbean.

Despite the great interest in international tourism demand and the growth in receipts, optimization of the tourism industry in the Caribbean Basin is a pressing matter. The region is losing market share, and the expenditure per visitor appears to be declining over time (WTO, 2005; Bolaky, 2011; Lorde, Li and Airey, 2016). This precarious situation is made worse by the fast-changing international tourism environment. Developing countries are no longer the only source markets destined for the Caribbean, but are, indeed, increasingly competing for tourists themselves. The growing technological innovations in tourism marketing and the ability of developing countries to master these technologies more rapidly could jeopardize the viability of the tourism industry in the region with its already meagre resources (Marius, 2015) with which to sustain its tourist market. Overall, the region has not been responding that well to the changing conditions in the international market spurred by globalization.

The region is also still reeling from the effects of the Great Recession, which stalled tourist arrivals, deteriorated economic imbalances, caused a decline in room capacity and imposed fewer flights to the Caribbean region (Laframbroise, Mwase, Park and Zhou, 2014). The productivity level in the region is lagging, which is impacting the quality of life of the population. The business climate in the region is affecting opportunities for growth, jobs and revenues (IMF, 2013). For example, islands in the Caribbean are placed in troubling ranking positions within the World Bank 'Doing Business' ranking (World Bank, 2016). On top of challenges created by the rapid process of globalization is the spectre of the impact for the individual countries in the region vis-à-vis the eventual reintegration of Cuba in the North American orbit. Studies such as Romeu (2008), Padilla and McElroy (2007) and Guzman (2015) allege that Cuba's presence in the US market would negatively impact the tourism prospects of several Caribbean islands.

It is the contention of this chapter that increased knowledge of the factors affecting the Caribbean region's tourism demand will enhance its tourism performance and thus could improve its growth prospects.

4.3 The Need to Improve Knowledge

Unquestionably, tourism demand research has grown in importance across the region and the world over the past two decades (Peng, Song, Crouch and Witt, 2015). Despite the rapid growth of international tourism and its importance in many countries, surprisingly little quantitative analysis has been conducted about small, warm-water destinations. These studies were conducted for the most part in developed countries, and they suggest that countries can benefit from rising real income and that the effect of changes in relative prices and exchange rates varies between countries, indicating that some countries are more sensitive to deteriorating price competitiveness than others.

A search from 2000 to 2017 has indicated that tourism demand is an under-researched area in Caribbean tourism. Only 21 studies related to demand analysis were published in academic journals during this time span. This confirms a similar finding of Shareef, Hoti and McAleer (2008), who found little research into tourism demand regarding small island destinations. Aside from the traditional topics related to demand determinants, forecasting and tourism demand modelling, topics that include climate change, business cycles and money supply are emerging as relevant areas in the tourism demand field. Table 4.1 reveals a selective list of journal publications about this topic referencing the Caribbean region.

Demand as defined by the willingness and ability to buy goods and services is the fundamental gauge of attractiveness of a tourist's destination. The demand for tourism is influenced by the economic concept of utility, i.e. the benefit or satisfaction that a person derives from the consumption of a good or service. Understanding demand requires a grasp of what comprises demand, the factors affecting the level of demand and insights of how future demand can be identified and estimated. These estimates are important to planning future tourism development, be it the government official engaged in maintaining the adequate level and proper quality of infrastructure, business climate and quality of life, or the private investor in search of a business opportunity and proposition.

Table 4.1. Selective journal publications about tourism demand in the Caribbean area.

Authors	Journal	Research focus
Moore (2010)	*Current Issues in Tourism*	Demand climate change
Greenidge (2001)	*Annals of Tourism Research*	Demand forecasting
Schubert, Brida and Risso (2011)	*Tourism Management*	Demand and economic growth
Levantis and Gani (2000)	*International Journal of Social Economics*	Demand and crime
Maloney and Rojas (2005)	*Applied Economics Letters*	Demand determinants
Pentelow and Scott (2011)	*Journal of Air Transport Management*	Demand and climate change policies
Padilla and McElroy (2007)	*Annals of Tourism Research*	Demand and Cuba
Bresson and Logossah (2011)	*Tourism Economics*	Demand and cruise tourism
Ridderstaat, Oduber, Croes, Nijkamp and Martens (2014)	*Tourism Management*	Demand and climate
Vanegas (2009)	*International Journal of Tourism Research*	Demand South American countries
Vanegas and Croes (2000)	*Annals of Tourism Research*	Demand determinants and elasticities
Pentelow and Scott (2010)	*Tourism and Hospitality Planning & Development*	Demand and climate change
Croes and Vanegas (2005)	*Tourism Management*	Demand determinants
Jackman and Greenidge (2010)	*Tourism and Hospitality Research*	Demand forecasting
Buckley (2011)	*AMBIO: A Journal of the Human Environment*	Demand and climate change
Korstanje and Clayton (2012)	*Worldwide Hospitality and Tourism Themes*	Demand and terrorism
Huybers and Bennett (2003)	*International Journal of Tourism Research*	Demand determinants
Ridderstaat and Croes (2015)	*Journal of Travel Research*	Demand and money supply
Mwase (2013)	*Applied Economics Letters*	Demand determinants
Onafowora and Owoye (2012)	*Tourism Economics*	Demand modelling
Shareef and McAleer (2005)	*International Journal of Tourism Research*	Demand modelling

Tourism demand research is also invaluable for marketing and promotional campaigns by keeping a destination or business in touch with its markets. Moreover, it can lead to the discovery of new markets, new products and more efficient uses of current products. Tourism demand analysis could also reduce the risk of unanticipated changes at the destination via the use of probable scenarios and alternative strategies. Improved knowledge of the factors contributing to the influx of tourist arrivals and tourist receipts is needed in the Caribbean region, especially when the prioritization of the allocation of scarce resources is required in development planning.

Elasticity provides an easily understandable basis of the variables examined by estimating the ratio of percentage changes or standard deviations (Song, Witt and Li, 2009; Croes, 2010). By understanding the variables that have an impact on demand, we have a greater ability to define what may happen in the future. Basing decisions on the elasticity estimation can be a helpful exercise in the set-up of a pricing strategy. For example, the value of the price elasticity can give the directional shift of total tourism revenue. When tourism demand is price elastic, raising the price will spawn a decrease in total tourism revenue; alternatively, when tourism demand is price inelastic, an increase in price will lead to an increase in total tourism revenue. Changing prices is not an easy exercise. Setting the optimal price has become a strategic advantage of any destination in the marketplace, particularly if it can avoid discouraging potential customers (Song and Wong, 2003). By creating this strategic capability, a destination avoids being consistently behind the curve on pricing and on missing important opportunities to maximize revenues.

These missed opportunities may explain the concerns expressed by some scholars related to the 'negative externalities' present in the industry today, such as devaluation of the destination, unemployment and underemployment, growing presence of prostitution and drugs, increased violence and ecological imbalances. In addition, there exist doubts as to whether tourism has the development prowess to increase and sustain growth over time. Some studies have found that tourism may be reacting to the Dutch disease or beach disease[2] revealing declining welfare in the long run. Evidence of this disease was found in the Balearics and Canary Islands (Capó, Font and Nadal, 2007). Alternatively, Holzner (2011) did not find evidence of the beach disease in his study of 130 countries. Patently, his study revealed that those countries that boasted higher income as a result of tourism also cultivated economic growth. Importantly, Holzner noted these same countries enjoyed the consequence of that income and economic growth rate in increased investment levels and increased higher education admissions. Further, there appeared low levels of distortion in the real exchange rates of tourism-dependent countries, thus indicating a greater balance in their outward orientation and resulting economic well-being. Finally, Holzner indicated that the industrial sectors of these countries seemed not to be impacted by retrenchment consequential to tourism growth.

The policymaker faced with these pressing problems may be tempted to leap to the 'easy' answer of increasing tourism in order to increase foreign exchange and decrease unemployment. However, failure to do so in a controlled manner may cause distortions in growth and result in damage to the ecology, excess carrying capacity and overuse of public utilities. In addition, tourism marketing organizations are increasingly asked to account for the effectiveness of marketing programmes through objective measurement that shows specific economic and social impacts, such as job creation, tax revenues and investments. Typically, arguments for allocating greater resources to the tourist sector in order to increase the levels of income have not been based on the existence of empirical analyses.

4.4 The Changing Competitive Setting

The demand paradigm assumes a rational individual is engaged in cost–benefit analysis when making a destination choice. The assumption gives rise to questions as the individual becomes involved in various vacation management plans. How are cost–benefit decisions reflected in a person's choice when shaping a vacation? How long does one work to gain the discretionary income for a vacation? How much leisure time does one sacrifice to increase that income? How much time and money should one spend on vacation? How does one allocate their financial resources among available goods and services? Where should one go? What transportation should one use? Where should one stay?

Mainstream tourism demand is characterized as a function of income, price and time. All three constructs have undergone important changes over time potentially affecting the capability of warm-water island destinations in their pursuit of optimizing their tourism industry. Several studies are claiming that the impact of income, while still important, is confounded by other factors. This claim is anchored in the observation that income elasticity has been decreasing over time, which means that tourism is losing its 'luxury' lustre. Peng *et al.* (2015) claim that income elasticity does not remain stable over time, due to the effects of economic fluctuations that affect expectations regarding income and employment. Smeral (2016) also asserts that income elasticity is not stable over time due to uncertainty induced by economic fluctuations. This uncertainty generates varying demand elasticities and can even reduce the income elasticity to less than one, thereby changing the tourism status from a luxury good to a normal good.

The decrease in elasticity is due to the increase in destination choices, money illusion and human psychology. Money illusion and human psychology influence how income is consumed thereby spawning business cycles. As demand flows transform and become transformed by business cycles and events, capacity utilization could also adjust via negative or positive alterations in tourism quality of service, opportunities for market share, industry profits, employment and the overall welfare of the host country and its residents. In terms of the Caribbean, Laframbroise *et al.* (2014) state that rapid financial effects are experienced in the tourism industry when the business cycles of more advanced economies adjust in ways that impact potential tourist abilities to travel. Thus, a marked macroeconomic incongruity is then transferred to the Caribbean.

Consequently, business cycles have become a variable in their own right to study tourism demand (Gouveia and Rodrigues, 2005; Guizzardi and Mazzocchi, 2010; Smeral, 2012, 2016; Merida and Golpe, 2016). These business cycles seem to have more explanatory power and seem more meaningful for tourism demand analysis. They reveal symmetric and asymmetric dynamics on tourism demand, which may reveal more insightful and meaningful information on which to base marketing decisions (Bronner and De Hoog, 2016).

However, Ridderstaat and Croes (2015) articulated two arguments that question the usefulness of business cycles as variable in tourism demand analysis. The critique is anchored on the timeliness of available data on the GDP (annual/quarterly basis) and the growing influence of the service sector in the GDP. This growing influence of the service sector mitigates the significance of industrial production, which conventionally has been used as a proxy for the GDP in business cycle analysis. In a Caribbean setting, business cycle analysis as a forward-looking (smart) tool becomes difficult due to the two points cited above. Money supply cycles in countries of origin may provide an interesting outlet for this forward-looking purpose.

Arguably, warm-water island destinations have little influence in shaping income outcomes in source countries. The region has little control over international tourism demand to the extent that demand is influenced by the rising wealth in the originating countries. These island destinations could respond effectively, however, and exert some control over tourism demand by gearing their marketing efforts towards intelligent pricing strategies. To enable the tourism industry to compete effectively, a better understanding of the income, price and exchange rate elasticities is required. In a competitive market, such as the Caribbean where more or less different islands could be considered substitutes for one another, if any destination raises its tourist prices, it may lose out unless it is unique, according to mainstream economics. For example, knowledge of the price elasticity of tourism market demand could play an important role in designing the appropriate strategy to realize more benefits from tourism.

Smeral (2003) revealed some structural issues that may impact price elasticity that, if not understood, could cloud tourism potential. Tourism seems to grow slower than the manufacturing industry over time because technological change seems faster in knowledge-intensive manufacturing than tourism, thereby impacting negatively on tourism cost structure. This means that the production costs of tourism will continue to grow over time, thus inducing higher prices with the potential for a drop in demand. While it seems difficult to attain cost advantage in tourism production, the only way to gain cost advantage seems to be on the demand side by making the tourism product more appealing, subsequently extracting higher prices over time (Keane, 1997; Croes, 2011). Working the tastes and preferences of tourists in such a way that tourists' goods are increasingly valued in international markets thereby offsetting the increase in cost production to raise or maintain prices of tourism products under conditions of increased competition seems a daunting task. The distinguished features of tourism, therefore, include a steady increase of production costs over time as compared with other economic sectors, and the opportunity to apply scarcity as a method to increase prices.

One reason why income and price have diminished in their information capability is due to changes in tastes and preferences. The opportunity to apply scarcity as a pricing strategy hinges on the proper knowledge of how tastes and preferences may affect tourism demand over time. Linder (1961) asserts that tourism flows between two countries are sparked by similar tastes and preferences, which is assumed to be revealed through common income level. Linder assumed that the income level (GDP per capita) provides a reasonable proxy to measure tastes. Buyers in higher income per capita countries demand higher quality goods, while those in relatively low income per capita countries demand lower quality goods. The more similar the demand for the products supplied by different countries, the greater the likelihood of trade between them. Consequently, there should be a clear association between international tourism and the level of GDP per capita.

Tastes and preferences are also shaped by the concept of time. The concept of time has become an important game changer in the tourist landscape. Veblen (1899), in his conspicuous behaviour study, alluded to the consumption of time only by the wealthy classes, while the other classes used time to recharge and re-energize to work. With increasing productivity, Keynes (1930) in his work *Economic Possibilities for our Grandchildren* predicted that leisure time would be in abundance to the point that his concern was how that generation would occupy itself with abundant leisure time in a meaningful way. Clearly, Keynes's prediction did not materialize in the absence of, or delay in, a leisure society. Commercialization of leisure and hyperactive lifestyles has consumed time and has fragmented society (Richards, 2010).

Changes in the amount of leisure time and differences between countries represent one of the more important aspects of the situation in the demand behaviour of tourists. For example, more time is spent on leisure in Germany, France and the UK than in the US (Stiglitz *et al.*, 2009). In the US, people tend to work more in order to consume more instead of enjoying more free time (Richards, 2010), while more people tend to live alone as solo dwellers (Klinenberg, 2012). According to the latter study, in the US more people are settling down as singletons; 27% of the population live alone with women leading in this statistic. On the other hand, tourism activities seem to be evolving from a relax-and-escape activity towards a relational activity, where people can meet each other to give meaning to their everyday life (Richards, 2010). These changing tastes and preferences have a profound impact on the way tourism is designed, produced and practised.

Pricing is predicated on adequate accessibility to an island destination. Islands face very different issues than mainland countries (Conlin and Baum, 1995; Yang, Mueller and Croes, 2016; Lorde *et al.*, 2016). Islands have natural barriers imposed by geography, distance and time. Almost any visitor who wants to stay overnight on an island would need air travel to get to the island, which means that air travel cost is one of the most important considerations in characterizing a destination choice. Regardless of its attractions, an island without proper accessibility, timely scheduling and reasonable prices would see its tourism industry imperilled and in danger of competitiveness erosion. Although the inclusion of transport costs is widely accepted in the tourism demand literature, many studies have excluded these costs from empirical studies due to multicollinearity or lack of data (Song and Wong, 2003).

These theoretical considerations are amplified in the Caribbean region by the potentially complete opening of Cuba to American tourists. The recent acceleration of detente between the US and Cuba has meant the easing of travel and trade restrictions on Cuba. Consequently, commercial US airlines began offering service between the countries for the first time in more than 50 years. Concerns have been expressed in the Caribbean with regard to the competitive pressure to regional destinations that would ensue from the complete liberalization of the relationship between the US and Cuba (Guzman, 2015). With restrictive lifts on US–Cuba relationships, the region's tourism market would need to re-examine and redefine its status quo in the tourism market. The results of potential entry of the Cuban market into Caribbean regional share has indeed forced a new consideration of tourism's role in upholding pricing strategies, base costs and tourism market consumption from egress countries to Caribbean destinations. With the potential for an emerging Cuban market, the Caribbean region could be forced to seek considerable redress of its market strategies (Romeu, 2008).

In conclusion, the conventional way of examining income, price and time as determinants of tourism demand, is shifting towards a more nuanced focus on tourism demand. The prospect of Cuba's complete reinsertion into the Caribbean tourist landscape adds to the degree of uncertainty regarding tourist flows to the other destinations in the region. Decreasing income elasticities, higher costs due to higher tourism production costs and changing tastes and preferences due to the dynamics of post modernism forces are compelling for a re-examining of how we look at tourism demand to warm-water island destinations.

4.5 The impact of Cuba on US tourism demand for a selected number of Caribbean countries: A cross price elasticity of demand analysis

As of 2016, travel restrictions for US nationals were loosened, and commercial US airliners were allowed to offer services to Cuba. This easing of constraints could have a ripple effect on those islands close to Cuba (Croes, 2006), while for other destinations the outcome would be muted in the end, according to the Caribbean Hotel & Tourism Association (2015). According to Romeu (2008), the opening of Cuba could cause an increase in overall arrivals to the Caribbean; other Caribbean islands would suffer losses from the US market, but then again are likely to compensate them with gains from capturing displaced tourists from other countries of origin who previously visited Cuba. A more recent study by Acevedo *et al.* (2016) found no negative impact from expanded flights to Cuba on the number of flights to the other islands of the Caribbean. The effects of a more actively participating Cuba in the US tourism demand pool are, thus, still subject to debate. The purpose of the ensuing analysis is to contribute to this discussion by determining the cross-price elasticity of US demand for a selected number of Caribbean countries (Aruba, Barbados, the Bahamas, Jamaica and the Dominican Republic), following price changes in Cuba.

The following tourism demand model was applied per destination in the analysis:

$$D_t = I_t + P_{D,t} + P_{C,t} + D_t + \varepsilon_t$$

Where:

D = US tourism demand for a destination involved in the analysis

I = per capita gross domestic product of the US (2009 = 100) (a proxy for income)

P_D = the relative price of the destination and the US (calculated as the ratio of the domestic price developments and the price developments in the US: adjusted for exchange rate differences)

P_C = the relative price of Cuba and the US (calculated as the ratio of the domestic price developments and the price developments in the US: adjusted for exchange rate differences)

D = dummy reflecting economic conditions (recession/expansion) in the US

t = time

ε = error term

Initially, the model also included a variable for transportation costs, but this was eliminated from the analysis due to multicollinearity. The annual data (2000–2016) was collected from several sources, including the central banks of Aruba, Barbados and the Dominican Republic; the tourism boards of the Bahamas and Jamaica; the Organization for Economic Co-operation and Development (an intergovernmental economic organization with 35 member countries aimed at promoting policies that will improve the economic and social well-being of people around the world); the Federal Reserve Economic Data (FRED), a database run by the Federal Reserve Bank of St. Louis; as well as from the calculations of the authors.

The data were subsequently decomposed in a trend and a cyclical component using the Hodrick-Prescott method. In this way, the analysis can distinguish between long-term (trend) and short-term (cycle) effects. After conversion to standardized values, both the cyclical and trend data were subjected to unit root tests (ADF, PP and KPSS), and the results indicated stationarity at both the level and first difference form. The authors chose to continue the analysis using the first difference form of the data, as first differencing is often sufficient to transform a time series into a stationary one (Jebb et al., 2015).

The cross-price elasticity of US demand for each island was estimated using a Two-Stage Least Squares method, which is a regression technique that controls for the possibility that one or more independent variables would be correlated with the error term in the regression. Using this technique would allow for the coefficients of the independent variables to be unbiased (i.e. uncorrelated with the error term). The results were also subjected to tests for model under-identification, endogeneity, serial correlation and heteroscedasticity, and the results were satisfactory. Before interpreting the results, one should understand that the short- and long-term effects should not be interpreted in a chronological sense (that is, the long-run effects should follow the short-run effects), but rather that both developments occur simultaneously, and the short-run effects should be interpreted as deviations from the long-run effects (Ridderstaat, Croes and Nijkamp, 2016b).

The results are provided in Table 4.2. In the case of Aruba, the long-term results indicate that only income (GDP per capita) is statistically relevant, but the relative price changes in Cuba (vis-à-vis the United States) are limited and statistically insignificant. The model results show that the effect of relative price changes in Cuba on US tourism demand for Aruba is only on the short run (10% significance). In other words, the negative statistically significant distortions caused by Cuban price developments are only temporary.

In the case of the Bahamas, only income was statistically significant in the short run, but changes in Cuban prices were not, probably because the Bahamas is relatively closer to the US than Cuba. The air travel distance between Miami and Nassau (the Bahamas) is about 182 miles, while that between Miami and Havana (Cuba) is 228 miles, which could be an important determinant of travel costs to both destinations. In an economic sense, the effects of price changes in Cuba on US tourism demand for the Bahamas is positive, both in the long and short term, indicating the presence of limited substitution effects between both markets.

With respect to Barbados, the results show that income was statistically relevant in both the long and short term. Moreover, the coefficients of price developments in Barbados vis-à-vis the US were positive in both the long and short haul (although only the prior was statistically significant at 5%), indicating the characteristics of a Veblen good[3] in Barbados' tourism. When considering the relative prices of

Table 4.2. 2SLS results (long and short term).

	I	$P_{D, AUA}$	$P_{D, BAH}$	$P_{D, BAR}$	$P_{D, JAM}$	$P_{D, DOM}$	P_C	D
Aruba								
Long-term	0.7731***	−0.0698					−0.0034	0.1129
Short-term	0.1003	−0.1052					−0.5121	0.6196
Bahamas								
Long-term	−0.0124		−0.2579				0.2757	−0.0949
Short-term	0.9768**		0.1709				0.1805	0.4382
Barbados								
Long-term	0.7516***			0.3555**			0.3510**	0.0415
Short-term	0.7952**			0.1791			−0.0305	0.6707
Jamaica								
Long-term	0.1288				0.9788***		−0.1836***	−0.0404
Short-term	0.4268				−0.1211		−0.7684	0.4325**
Dominican Republic								
Long-term	0.6542***					−0.4036***	0.4358**	0.0856
Short-term	−0.2579					−1.1898***	−0.6236*	0.2921

Note: ***, ** and *indicate significance at, respectively, 1%, 5% and 10%. AUA = Aruba; BAH = Bahamas; BAR = Barbados; JAM = Jamaica; DOM = Dominican Republic

Cuba vis-à-vis those in the US, the results show a slight temporary negative distortion (although statistically insignificant), but in the long run there could be substitution effects in the demand for tourists from these two countries.

In the case of Jamaica, relative price changes vis-à-vis the US could provide a temporary negative alteration in US tourism demand for this destination (although not statistically significant), but the long-run effects indicate that US tourism demand for Jamaica has the features of a Veblen good. When considering price changes in Cuba vis-à-vis the US, the effect was negative (both short- and long-term), indicating that, under the assumption of the Veblen physiognomies of US tourism demand for Jamaica, there could be substitution effects between both destinations. Economic conditions also affect tourism demand for Jamaica, particularly in the short run, whereby short-term periods of recession are likely to have a positive effect on US tourism demand, possibly indicating that US visitors are likely to choose short-haul visits over long-haul ones in times of economic crisis (Smeral, 2009). In the long run, the effect of economic conditions on US tourism demand for Jamaica is limited.

The case of the Dominican Republic shows that income is a statistically relevant factor in the long run (short run: slightly negative, but statistically irrelevant). Developments in domestic prices vis-à-vis those in the US are statistically significant and negative in both the long and short term, indicating that tourism in the Dominican Republic is a substitute good. Moreover, price developments in Cuba have a negative effect in the short term, indicating a temporary negative distortion on US tourism demand for the Dominican Republic. However, in the long run, the effect is positive and also statistically significant, indicating a cross-substitution effect between both destinations.

Overall, the results show that price dynamics in Cuba are likely to have ambiguous effects, depending on the island in the analysis, as well as the considered perspective (long run versus short run). Furthermore, the findings provide an invitation for further analysis of Cuba's developments on other islands in the region in the near term.

4.6 Emerging Drivers of Tourism Demand

Literature has indicated tourism demand (and thus development) to be dynamic in nature (McKercher, 1999; Russel and Faulkner, 1999; Baggio, 2008). Moreover, this nature may be explained by numerous influences that range from behavioural changes and recurring or cyclical variations, to unforeseen events (Ridderstaat, 2015). Yet, while researchers have acknowledged the vigorous nature of tourism demand, an understanding of the driving forces behind that demand phenomenon has largely remained stationary. Also, researchers may not have considered all explanatory variables in each individual study, the result of which would further compromise a complete understanding of the driving forces. Three problems that have been encountered in numerous demand studies (Ridderstaat, 2015) might explain this stationarity. First, tourism demand research has, for a long time, been structured around a small selected number of explanatory variables, such as income, prices (including substitute prices), exchange rates, transportation costs, marketing expenses, population of source countries, expectations, habit persistence and one-off effects (e.g. oil prices) (Song, Witt and Lee, 2009; Croes, 2010; Goh, 2012). Second, tourism demand studies have, for a long time, been based on annual data, which hide potential influences occurring at more frequent periodicities, such as on a monthly or quarterly basis. Third, tourism demand studies have been largely oblivious regarding the bilateral effects involving tourism demand, particularly from the perspective of considering the manner in which tourism demand is influenced by the consequences of the very factors that it promulgates. For instance, while studies have amply considered the effect of tourism on the economy, only a select number of these investigations have explored the reverse relationship (Ridderstaat, Croes and Nijkamp, 2014). In recent years, a number of supplementary drivers of tourism demand have been proposed, in addition to the conventional drivers.

Climate

Climate, as an explanatory factor of tourism demand, has been investigated both from the perspective of climate change (Scott, McBoyle and Schwartzentruber, 2004; Hamilton, Maddison and Tol, 2005; Amelung, Nicholls and Viner, 2007; Hamilton and Tol, 2007; Kulendran and Dwyer, 2010) and periodical changes in weather (e.g. Scott and McBoyle, 2001; Yu, Schwartz and Walsh, 2009, 2010; Ridderstaat et al., 2014). Climate change studies have covered the long-term effects of climate on tourism demand (e.g. Behringer, Buerki and Fuhrer, 2000; Scott *et al.*, 2004; Hamilton, Maddison and Tol, 2005; Berrittella *et al.*, 2006; Hamilton and Tol, 2007; Soboll and Dingeldey, 2012). On the other hand, studies on periodical changes of weather factors have focused on the short-term effects on tourism demand, like the seasonal perspective (e.g. Yu, Schwartz and Walsh, 2009, 2010; Ridderstaat *et al.*, 2014). The latter studies required in many instances a decomposition of time series data.

Business cycles

Studies have also covered the impact of business cycles on tourism demand flows (e.g. Gouveia and Rodrigues, 2005; Guizzardi and Mazzocchi, 2010; Narayan, 2011; Smeral, 2012; Smeral and Song, 2015; Bronner and de Hoog, 2016; Merida and Golpe, 2016). Often, these studies distilled the cyclical factor from the GDP series, but the use of this variable is hampered by a lack of data on a periodicity lower than the quarterly basis. Researchers have tried to use monthly industrial production indices as a proxy for the GDP, but the use of this variable becomes complicated as economies are becoming more and more dependent on the service sector (Fulop and Gyomai, 2012). The challenge is then to look for alternative proxies of business cycle activities, for example, money supply.

Money supply

Money supply has been brought forward as a determinant of tourism demand, although studies in this area are still in the infant stages. As an indicator, money supply data are, generally, easier to compile, have a lower production lag, and are published on a much more frequent basis than the GDP (Ridderstaat and Croes, 2015). This makes this variable particularly attractive as a forerunner of tourism demand development. Only a few studies have considered money supply as a potential driver of tourism demand (Turner, Kulendran and Fernando 1997; Cho 2001; Choi 2003, Ridderstaat and Croes, 2015). Particularly the cyclical factor of this variable seems to have a connection with the tourism demand cycle, although there is also room for researching the other elements of money supply (e.g. seasonal and trend factors).

Quality of life

Many studies have looked at the effects of tourism development on quality of life of destinations' citizens (Andereck and Vogt, 2000; Kim, 2002; Fredline, Deery and Jago, 2005; Andereck *et al.*, 2007; Marzuki, 2009; Andereck and Nyaupane, 2010), but recent studies indicate the opposite relation could be conceivable as well. Quality of life could affect tourism demand development through four different mechanisms, i.e. through: (i) support for tourism; (ii) provision of amenities; (iii) education of citizens; and (iv) citizens' deliberate misusing of tourists for personal gains (Ridderstaat, Croes and Nijkamp, 2016a, 2016b). The number of studies investigating the impact of quality of life on tourism demand development is still small (Croes, 2012; Woo,

Kim and Uysal, 2015; Ridderstaat, Croes and Nijkamp, 2016a, 2016b) but offering an interesting and important avenue for further exploration.

Other drivers

A quite limited number of studies have focused on the effect of diseases on tourism demand, such as chikungunya and/or dengue (Mavalankar *et al.*, 2009; Oduber, Ridderstaat and Martens, 2014), or on the effects of vegetation issues (Oduber, Ridderstaat and Martens, 2015).

The findings of the studies above indicate that tourism demand is a complex phenomenon that requires a determined inquiry into its causal factors.

4.7 A Strategy for What is Yet to Come

It is clear that tourism demand is the outcome of a scaling number of driving forces, which has implications for our understanding of its impending evolution. The quest is to continue to find the missing pieces to solve this big puzzle in tourism. The specific challenge for each warm-water destination is, like Sherlock Holmes, to see what others do not see, and to act proactively, rather than to respond reactively to developments.

One research strategy is to try to connect pieces that are recognizable to the conventional pieces: to look for connectors between the already known drivers and other data that could come available in a timely manner. For instance, while it is generally known that economic development in countries of origin could be an influencing factor in tourism demand changes, countries' GDP data are usually available with a lag. So, the question here is what other indicators could provide us with a leading indication of economic activity in the source countries. Initial results indicate, for example, that money supply could provide an early understanding of the approaching dynamics of tourism demand; so further research in this area is a worthy endeavour. Also, researching other monetary aggregates (for example currency in circulation and deposits by duration), as to whether they have a connection with tourism demand, should be part of the approach.

A second strategy is to look for random connectors of tourism demand, i.e. those factors that seemingly have no connection with conventional determinants of tourism demand. Climate and quality of life of residents are good examples of these random drivers, but the area of diseases also provides an interesting avenue to explore. The mission here is also to investigate the elements within these random determinants that could provide a leading indicator of tourism development, thereby contributing further to anticipation. For instance, quality of life of residents consists of many dimensions (e.g. education, income, social participation), each of which could provide an early warning on impending changes on the overall quality of life and, subsequently, tourism demand.

In the spirit of Sherlock Holmes, one important issue is not only to find the malefactors of tourism demand, but also to take the right actions afterwards. Discovering new determinants for tourism demand is only part of the whole expedition. One must also be able to understand what types of policy measures are needed (if they are needed!), and in what quantity. Policy application in tourism demand management

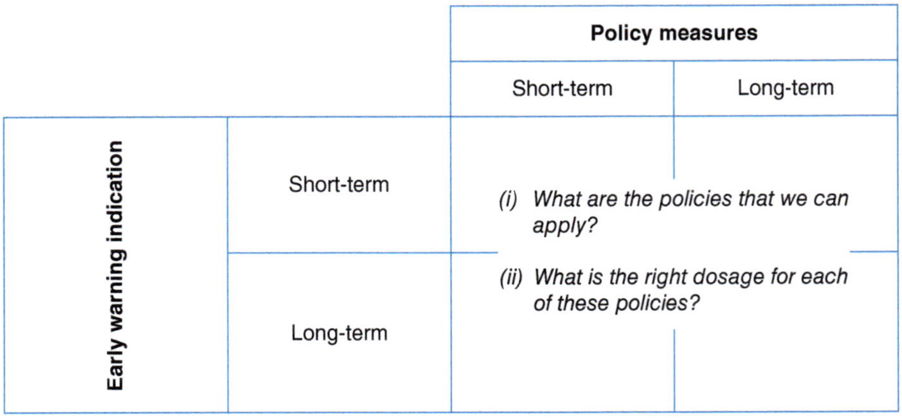

Fig. 4.1. A research framework for a pro-active tourism demand managing toolbox.

requires discovery of methods to counter or stimulate anticipated developments in tourism in order to best reap the benefits of tourism development.

Figure 4.1 proposes a study framework for a two-way approach to tourism demand management, which includes both early warning indicators and associated policy measures. The figure includes a further dimension by distinguishing between both the short-term and long-term phases. It is important to discriminate between indicators anticipating short- or long-term effects on tourism demand, given that the previous involves deviations from the latter and, as such, both may require different policy prescriptions to effectively manage tourism demand developments.

Tourism is expected to be the continued driver of small islands' economies, at least in the coming decades. The quest for each of these warm-water destinations is to optimally profit from this phenomenon by outwitting the competition and winning the competitive superiority challenge.

Notes

[1] The term Dutch disease references tourism expansion to the detriment of other economic sectors, causing welfare reduction in a society. Tourism expansion may shift resources towards non-tradable sectors in the economy at the expense of productivity loss and may spawn a persistent real exchange rate appreciation affecting the economic growth in the long term.

[2] Beach disease is a term coined by Holzner (2011) and refers to the Dutch disease phenomenon affecting tourism-dependent countries in the long run.

[3] A Veblen good refers to goods (or services) that are purchased by users to create or elevate their status or image (e.g. expensive wines, jewellery or luxury holidays). The demand for a Veblen good increases with rising prices for the specific product or service

References

Acevedo, S., Han, L., Kim, H. and Laframboise, N. (2016) *Flying to Paradise; The Role of Airlift in the Caribbean Tourism Industry* (No. 16/33). International Monetary Fund.

Amelung, B., Nicholls, S. and Viner, D. (2007) Implications of global climate change for tourism flows and seasonality. *Journal of Travel Research* 45, 285–296.

Andereck, J. and Nyaupane, G. (2010) Exploring the nature of tourism and quality of life perceptions among residents. *Journal of Travel Research* 50(3), 248–60.

Andereck, K.L. and Vogt, C.A. (2000) The relationship between residents' attitudes toward tourism and tourism development options. *Journal of Travel Research* 39(1), 27–36.

Andereck, K., Valentine, K., Chr. Vogt and Vogt, V. (2007) A cross-cultural analysis of tourism and quality of life perceptions. *Journal of Sustainable Tourism* 15(5), 483–502.

Baggio, R. (2008) Symptoms of complexity in a tourism system. *Tourism Analysis* 13(1), 1–20.

Behringer, J., Buerki, R. and Fuhrer, J. (2000) Participatory integrated assessment of adaptation to climate change in Alpine tourism and mountain agriculture. *Integrated Assesment* 1(4), 331–338.

Berrittella, M., Bigano, A., Rosen, R. and Tol, R. (2006) A general equilibrium analysis of climate change impacts on tourism. *Tourism Management* 27, 913–924.

Bolaky, B. (2011) Tourism competitiveness in the Caribbean. *Cepal Review* 104, 55–76.

Brau, R., Lanza, A. and Pigliaru, F. (2007) How fast are small tourism countries growing? Evidence from the data for 1980–2003. *Tourism Economics* 13, 603–613.

Bresson, G. and Logossah, K. (2011) Crowding-out effects of cruise tourism on stay-over tourism in the Caribbean: Non-parametric panel data evidence. *Tourism Economics* 17(1), 127–158.

Brida, J., Cortes-Jimenez, I. and Pulina, M. (2016) Has the tourism-led growth hypothesis been validated? A literature review. *Current Issues in Tourism* 19(5), 394–430.

Bronner, F. and De Hoog, R. (2016) Tourist demand reactions: Symmetric or asymmetric across the business cycle? *Journal of Travel Research* DOI:10.1177/0047287516672347, 1–15.

Buckley, R. (2011) Tourism under climate change: Will slow travel supersede short breaks? *AMBIO: A Journal of the Human Environment* 40(3), 328–331.

Buswell, R. (2011) *Mallorca and tourism: history, economy, and environment*. Channel View Publications, Bristol, UK.

Capó, J., Font, A. and Nadal, J. (2007) Dutch disease in tourism economies. Evidence from the Balearics and the Canary Islands. *Journal of Sustainable Tourism* 15(6), 615–627.

Caribbean Hotel & Tourism Association (2015) Cuba: The Great Disruption for the Good of the Caribbean. The Caribbean Hotel and Tourism Association states its views. Available at: https://ccaadcblog.files.wordpress.com/2015/06/chta-cuba-paper-final-6-18-15.pdf (accessed 9 October 2017).

Cho, V. (2001) Tourism forecasting and its relationship with leading economic indicators. *Journal of Hospitality & Tourism Research* 25(4), 399–420.

Choi, J. (2003) Developing an economic indicator system (a forecasting technique) for the hotel industry. *Hospitality Management* 22(2), 147–59.

Conlin, M. and Baum, T. (1995) *Island Tourism: Management Principles and Practices*. Wiley, Chichester, UK.

Croes, R. (2006) A paradigm shift to a new strategy for small island economies: embracing demand side economics for value enhancement and long term economic stability. *Tourism Management* 27(3), 453–465.

Croes, R. (2010) *Anatomy of Demand in International Tourism: The Case of Aruba*. Lambert Academic Publishing, Saarbrucken, Germany.

Croes, R. (2011) *The Small Island Paradox. Tourism Specialization as a Potential Solution*. Lambert Academic Publishing, Saarbrucken, Germany.

Croes, R. (2012) Assessing tourism development from Sen's capability approach. *Journal of Travel Research* 51(5), 542–554.

Croes, R. (2013) Tourism specialization and economic output in small island destinations. *Tourism Review* 68(4), 34–48.

Croes, R. and Vanegas, M. (2005) An econometric study of tourist arrivals in Aruba and its implications. *Tourism Management* 26(6), 879–890.

Fredline, L., Deery, M. and Jago, L. (2005) Social impacts of tourism on communities. Available at: http://www.surfcoast.vic.gov.au (accessed 9 December 2010).

Fulop, G. and Gyomai, G. (2012) Transition of the OECD CLI System to a GDP-based Business Cycle Target. Composite Leading Indicator for background note. OECD, Paris, France.

Goh, G. (2012) Exploring the impact of climate on tourism demand. *Annals of Tourism Research* 39(4), 1859–1883.

Gouveia, P. and Rodrigues, P. (2005) Dating and synchronizing tourism growth cycles. *Tourism Economics* 11(4), 501–515.

Greenidge, K. (2001) Forecasting tourism demand: An STM approach. *Annals of Tourism Research* 28(1), 98–112.

Guizzardi, A. and Mazzocchi, M. (2010) Tourism demand for Italy and the business cycle. *Tourism Management* 31(3), 367–377.

Guzmán, J.A.A. (2015) Perspectivas del sector hotelero en la Republica Dominicana a partir del desbloqueo cubano. *TURyDES* 8(18), 1–8.

Hamilton, J.M. and Tol, R.S. (2007) The impact of climate change on tourism in Germany, the UK and Ireland: a simulation study. *Regional Environmental Change* 7(3), 161–172.

Hamilton, J.M., Maddison, D.J. and Tol, R.S. (2005) Climate change and international tourism: a simulation study. *Global Environmental Change* 15(3), 253–266.

Holzner, M. (2011) Tourism and economic development: the beach disease? *Tourism Management* 32(4), 922–933.

Huybers, T. and Bennett, J. (2003) Environmental management and the competitiveness of nature-based tourism destinations. *Environmental and Resource Economics* 24, 213–233.

IMF (2013) Caribbean small states: challenges high debt and low growth. Available at: https://www.imf.org/external/np/pp/eng/2013/022013b.pdf (accessed 24 February 2017).

Jackman, M. and Greenidge, K. (2010) Modelling and forecasting tourist flows to Barbados using structural time series models. *Tourism and Hospitality Research* 10(1), 1–13.

Jebb, A., Tay, L., Wang, W. and Huang, Q. (2015) Time series analysis for psychological and forecasting change. *Frontiers in Psychology* 6, 1–24.

Keane, M. (1997) Quality and pricing in tourism destinations. *Annals of Tourism Research* 24(1), 117–130.

Kenyes, J. (1930) *Economic possibilities of our grandchildren*. Available at: http://www.econ.yale.edu/smith/econ116a/keynes1.pdf (accessed 24 February 2017).

Kim, K. (2002) The effects of tourism impacts upon quality of life of residents in the community. PhD dissertation. Virginia Polytechnic Institute and State University, Blacksburg, Virginia.

Klinenberg, E. (2012) *Going Solo. The extraordinary rise and surprising appeal of living alone*. The Penguin Press, New York.

Korstanje, M. and Clayton, A. (2012) Tourism and terrorism: conflicts and commonalities. *Worldwide Hospitality and Tourism Themes* 4(1), 8–25.

Kulendran, N. and Dwyer, L. (2010) *Seasonal Variation versus Climate Variation for Australian Tourism*. CRC for Sustainable Tourism Pty Limited, Queensland, Australia.

Laframboise, M., Mwase, N., Park, M. and Zhou, Y. (2014) *Revisiting Tourism Flows to the Caribbean: What is Driving Arrivals?* IMF Working Paper, WP/14/229, Washington, DC.

Lanza, A. and Pigliaru, F. (2000) Tourism and economic growth: does country's size matter? *Rivista Internazionale di Scienze Economiche e Commerciali* 47, 77–85.

Levantis, T. and Gani, A. (2000) Tourism demand and the nuisance of crime. *International Journal of Social Economics 27*, 959–967.

Linder, S. (1961) *An Essay on Trade and Transformation*. Wiley, New York.

Lorde, T., Li, G. and Airey, D. (2016) Modeling Caribbean tourism demand: an augmented gravity approach. *Journal of Travel Research* 55(7), 946–956.

Maloney, W.F. and Montes Rojas, G.V. (2005) How elastic are sea, sand and sun? Dynamic panel estimates of the demand for tourism. *Applied Economics Letters* 12(5), 277–280.

Marius, M. (2015) Snapshot: 2015 update of ICT network readiness in the Caribbean. Available at: http://www.ict-pulse.com/2015/04/snapshot-2015-update-ict-network-readiness-caribbean/ (accessed 24 February 2017)

Marzuki, A. (2009) Impacts of tourism development. *Anatolia* 20(2), 450–455.
Mavalankar, D., Puwar, T., Murtola, T. and Vasan, S. (2009) Quantifying the impact of Chikungunya and Dengue on tourism revenues. Working Paper No. 2009-02-03, Indian Institute of Management, Ahmedabad, India.
McKercher, B. (1999) A chaos approach to tourism. *Tourism Management* 20, 425–434.
Merida, A. and Golpe, A. (2016) Tourism-led growth revisited for Spain: Causality, business cycles and structural breaks. *International Journal of Tourism Research* 18, 39–51.
Moore, W. (2010) The impact of climate change on Caribbean tourism demand. *Current Issues in Tourism* 13(5), 495–505.
Mwase, N. (2013) Tourism flows to Caribbean islands: an empirical note. *Applied Economics Letters* 20(10), 957–965.
Narayan P (2011) Are shocks to tourism transitory at business cycle horizon? *Applied Economics* 43(16), 2071–2077.
Oduber, M., Ridderstaat, J. and Martens, P. (2014) The bilateral relationship between tourism and dengue occurrence: Evidence from Aruba. *Journal of Tourism and Hospitality Management* 2(6), 223–244.
Oduber, M., Ridderstaat, J. and Martens, P. (2015) The connection of vegetation with tourism development and economic growth: A case study for Aruba. *Journal of Environmental Science and Engineering* A 4, 420–431.
Onafowora, O. and Owoye, O. (2012) Modelling international tourism demand for the Caribbean. *Tourism Economics* 18(1), 159–180.
Padilla, A. and McElroy, J. (2007) Cuba and Caribbean tourism after Castro. *Annals of Tourism Research* 34(3), 649–672.
Peng, B., Song, H., Crouch, G. and Witt, S. (2015) A meta-analysis of international tourism demand elasticities. *Journal of Travel Research* 54(5), 611–633.
Pentelow, L. and Scott, D. (2010) The implications of climate change mitigation policy. Volatility of tourism arrivals to the Caribbean. *Tourism and Hospitality Planning and Development* 7, 301–315.
Pentelow, L. and Scott, D. (2011) Aviation's inclusion in international climate policy regimes: implications for the Caribbean tourism industry. *Journal of Air Transport Management* 17(3), 199–205.
Pratt, S. (2015) The economic impact of tourism in SIDS. *Annals of Tourism Research* 52, 148–160.
Richards, G. (2010) Leisure in the network society: from pseudo-events to hyperfestivity? Tilburg University, the Netherlands.
Ridderstaat, J. (2015) Studies on determinants of tourism demand dynamics in a small island destination, the case of Aruba. PhD thesis, Free University, Amsterdam, the Netherlands.
Ridderstaat, J. and Croes, R. (2015) The link between money supply and tourism demand cycles: A case study for two Caribbean destinations. *Journal of Travel Research* 56(2), 187–205.
Ridderstaat, J., Croes, R. and Nijkamp, P. (2014) Tourism and Long-run Economic Growth in Aruba. *International Journal of Tourism Research* 16(5), 472–487.
Ridderstaat, J., Oduber, M., Croes, R., Nijkamp, P. and Martens, P. (2014) Impacts of seasonal patterns of climate on recurrent fluctuations in tourism demand: Evidence from Aruba. *Tourism Management* 41, 245–256.
Ridderstaat, J., Croes, R. and Nijkamp, P. (2016a) A two-way causal chain between tourism development and quality of life in a small island destination: an empirical analysis. *Journal of Sustainable Tourism* 24(10), 1461–1479.
Ridderstaat, J., Croes, R. and Nijkamp, P. (2016b) The tourism development–quality of life nexus in a small island destination. *Journal of Travel Research* 55(1), 79–94.
Romeu, R. (2008) Vacation over: Implications for the Caribbean of opening U.S.–Cuba Tourism. International Monetary Fund Working Paper WP/08/162.
Russel, R. and Faulkner, B. (1999). Movers and shakers: chaos makers in tourism development. *Tourism Management* 20(4), 411–423.
Schubert, S., Brida, J. and Risso, W. (2011) The impacts of international tourism demand on economic growth of small economies dependent on tourism. *Tourism Management* 32(2), 377–385.

Scott, D. and McBoyle, G. (2001) Using a 'tourism climate index' to examine the implications of climate change for climate as a tourism resource. In: Matzarakis, A. and de Freitas, C. (eds) *Proceedings of the First International Workshop on Climate, Tourism and Recreation* (pp. 69–88). Available at: http://www.mif.uni-freiburg.de/ISB/ws/papers/full_report.pdf#page¼73 (accessed 17 July 2013).

Scott, D., McBoyle, G. and Schwartzentruber, M. (2004) Climate change and the distribution of climatic resources for tourism in North America. *Climate Research* 27(2), 105–117.

Seetanah, B. (2011) Assessing the dynamic economic impact of tourism for island economies. *Annals of Tourism Research* 38(1), 291–308.

Shareef, R. and McAleer, M. (2005) Modelling international tourism demand and volatility in small island tourism economies. *International Journal of Tourism Research* 7(6), 313–333.

Shareef, R., Hoti, S. and McAleer, M. (2008) *The Economics of Small Island Tourism. International Demand and Country Risk Analysis*. Edward Elgar, Cheltenham, UK.

Smeral, E. (2003) A structural view of tourism economics. *Tourism Economics* 9(1), 77–94.

Smeral, E. (2009). The impact of the financial and economic crisis on European tourism. *Journal of Travel Research* 48(1), 3–13.

Smeral, E. (2012) International tourism demand and the business cycle. *Annals of Tourism Research* 39(1), 379–400.

Smeral, E. (2016) Tourism forecasting performance considering the instability of demand elasticities. *Journal of Travel Research* DOI: 0047287516671435.

Smeral, E. and Song, H. (2015) Varying elasticities and forecasting performance. *International Journal of Tourism Research* 17, 140–150.

Soboll, A. and Dingeldey, A. (2012) The future impact of climate change on alpine winter tourism: a high-resolution simulation system in the German and Austrian Alps. *Journal of Sustainable Tourism* 20(1), 101–120.

Song, H. and Wong, K. (2003) Tourism demand modeling: a time- varying parameter approach. *Journal of Travel Research* 42(1), 57–64.

Song, H., Witt, S. and Li, G. (2009) *The Advanced Econometrics of Tourism Demand*. Routledge, Abingdon, UK.

Stiglitz, J., Sen, A. and Fitoussi, J. (2009) Report by the Commission on the Measurement of Economic Performance and Social Progress. Available at: http://library.bsl.org.au/jspui/bitstream/1/1267/1/Measurement_of_economic_performance_and_social_progress.pdf (accessed 23 February 2017).

Turner, L., Kulendran, N. and Fernando, H. (1997) The use of composite national indicators for tourism forecasting. *Tourism Economics* 3(4), 309–17.

Vanegas, M. (2009) Tourism demand response by residents of Latin American countries. *International Journal of Tourism Research* 11(1), 17–29.

Vanegas, M. and Croes, R. (2000) Evaluation of Demand: U.S. Tourists to Aruba, *Annals of Tourism Research* 27(4), 946–963.

Veblen, T. (1899) *The Theory of the Leisure Class: An Economic Study of Institutions*. Macmillan, New York.

Woo, E., Kim, H. and Uysal, M. (2015) Life satisfaction and support for tourism development. *Annals of Tourism Research* 50, 84–97.

World Bank (2016) Doing business. Available at: http://www.doingbusiness.org/rankings (accessed 24 February 2017).

WTO (2005) *A Time to Choose: Caribbean Development in the 21st century*. WTO, Washington, DC, April.

WTTC (2015) Economic Impact 2015 Caribbean. Available at: https://www.wttc.org/-/media/files/reports/economic%20impact%20research/regional%202015/caribbean2015.pdf (accessed 21 February 2017).

Yang, Y., Mueller, N. and Croes, R. (2016) Accessibility and hotel prices in the Caribbean: The moderating effect of quality-signaling factors. *Tourism Management* 56, 40–51.

Yu, G., Schwartz, Z. and Walsh, J. (2009) Effects of climate change on the seasonality of weather for tourism in Alaska. *Arctic* 62(4), 443–457.

Yu, G., Schwartz, Z. and Walsh, J. (2010) Climate change and tourism seasonality. *Journal of Tourism* 9(2), 51–65.

5 Island Hospitality, Services and Businesses

ANNE CRICK*

The University of the West Indies, Mona, Jamaica

5.1 The Changing Face of Caribbean Hospitality

Islands are seen as idyllic, an escape from reality and even as paradise (King, 1997), and every year attract millions of tourists to their shores. For almost a century tourism has helped Caribbean governments to bring in foreign direct investment, earn foreign exchange and create jobs for their citizens. Tourism has, through its multiplier effect, spawned new industries particularly in the area of hospitality. As tourism becomes a mature product it faces new challenges and must adjust its strategies (Croes, 2006). However as is always the case, there are opportunities that may be seized by bold and visionary entities.

Caribbean entrepreneurs face a number of challenges in trying to seize these opportunities. First, it may be difficult for small islands to distinguish themselves in a crowded marketplace, and each destination needs to find a way to attract and retain a steady stream of visitors. Second, the islands face the small island paradox in which the price of the tourism product tends to increase over time in older destinations, which then are confronted with competition from lower priced new destinations (Croes, 2011). Destinations must therefore continually refresh their products and services. Third, islands must balance the need to import goods and services that are appealing to an international market with the need to support local businesses and to reduce foreign exchange leakage. Fourth, islands must also balance the need to grow with the risks associated with advancement up the destination life cycle (McElroy, 2006). They must therefore ensure that the needs and concerns of the local population are balanced against the needs and concerns of visitors. Finally, islands must adapt to the needs of the 'new tourists' who are growing in number.

This chapter focuses on the emerging dynamic of both 'new tourists' and millennials and identifies the implications of these groups for small island hospitality

*E-mail: anne.crick@uwimona.edu.jm

businesses in light of the challenges they face. The chapter continues with a discussion of the implications of 'new tourism' for hospitality and contrasts it with 'old tourism', which is characterized by the tenets of McDonaldization.[1] This is followed by a discussion of the implications for the delivery of quality service and products to both sets of tourists. The roles for the public and private sector are then discussed and the chapter concludes with lessons learned and policy recommendations.

5.2 From Old to New Tourism: Implications for Hospitality

Hospitality has been defined as the provision of food and/or drink and/or accommodation away from home (The Joint Hospitality Industry Congress, 1996 as cited in Lashley, 2001). It therefore potentially includes every bar, cookshop, fruit stand, restaurant, hotel, guest house and resort on the island. It also incorporates formal and informal, large and small, commercial and domestic entities. The scope of Caribbean hospitality is therefore very broad. Hospitality and tourism are inextricably intertwined in the Caribbean and indeed many hospitality businesses exist only because of tourism. All of these businesses must meet international expectations, and they must do so in a way that caters to the needs of two distinct groups – 'old tourists' and 'new tourists'.

Old tourists

Much of Caribbean hospitality may be associated with old tourism, which is characterized by mass, standardized and rigidly packaged holidays, hotels and tourists (Poon, 1993). Old tourism provides consumers with predictability and security and it benefits hospitality businesses with more predictable guest behaviour. It may be analysed using the tenets of McDonaldization – *efficiency*, *calculability*, *predictability* and *control* (Ritzer, 1996).

Efficiency refers to the optimum method of performing any task and customers in pursuit of efficiency opt for the easiest and most convenient solutions. For example, all-inclusive hotels where all food, beverage, entertainment and most sports are included; branded multinational hospitality chains make decision making easier for customers because of the efficiency of their packaged solutions. These businesses may therefore enjoy an advantage with 'old tourists'. Hospitality organizations may also choose to benchmark international brands known for their efficiency, such as Trinidadian coffee chain Rituals has done with American coffee shops (Jacob, 2011).

Calculability focuses on things that may be calculated, counted or quantified with the implicit suggestion that more is better. All-inclusive chains such as Sandals with the tagline 'All included, All unlimited, All the time' (Sandals, 2017), best epitomize this tenet as do restaurants promising all you can eat buffets or bottomless mimosas. Economies of scale may make calculability feasible for larger entities but challenging for smaller ones.

Predictability refers to the removal of surprise from the service experience and may be attractive to 'old tourists' who are comfortable with packages and standardization. Again, chain-branded hospitality businesses and all-inclusive hotels reduce uncertainty and will be popular with these tourists. They also provide operators with greater certainty and in the case of chains, management support.

Control suggests that everything is pre-planned and measured according to a predetermined standard and formula. 'Old tourists may enjoy knowing exactly what to expect and familiar brands may help them to feel more in control of the situation, even when they are in unfamiliar territory. Control can also help businesses to better manage costs.

The demand by 'old tourists' for McDonaldized types of products and services helps to explain the structure and contours of Caribbean hospitality businesses. As tourism has become more embedded in destinations, branded and larger entities that meet the needs of 'old tourists' may have had an advantage over small, local establishments. All-inclusives that offer significant predictability and control have become increasingly popular, and many European Plan (EP) hotels have changed from room only to the all-inclusive model in order to be more competitive. While this type of hospitality business has been beneficial in meeting the needs of 'old tourists', there is a risk that Caribbean destinations may lose their uniqueness if they feature similar hotel, bar and restaurant brands and experiences across their landscapes.

There is also a fifth aspect of McDonaldization – irrationality, which suggests that taken to the extreme, the characteristics of McDonaldization become unreasonable. Control may, for example, make the customer experience more limiting and less unique and personalized. Predictability has its downside as well. For example, while it may be comforting to hear consistent phrases by employees working for a particular brand, it may also be jarring if it is not a fit with the local culture. McDonaldized products and services may also provide quantity at the expense of quality (Ritzer, 1996), and reduce loyalty and referrals. Finally, 'old tourism' is also likely to include staged authenticity, which gives a flavour of the destination in measured and controlled doses making it more palatable to 'old tourists'. In Jamaica for example, the Braco resort has recreated a colonial era town complete with a rum bar and cricket pitch onsite for guests. This only works however if it is convincingly authentic and it may be unacceptable to many tourists.

Will old tourism disappear? That is unlikely since it offers visitors some assurances when they travel overseas (McAlone, 2015), and risk-averse tourists will always enjoy knowing exactly how much their vacation will cost, and will appreciate the security of an all-inclusive hotel. 'New tourists' and millennials have different expectations and needs however. These are discussed in the next section.

The 'new tourists' and the millennials

'New tourism' by contrast, is characterized by flexibility and more authentic experiences. Therefore, while 'old tourists' would typically be part of a group and would follow the masses, 'new tourists' would be individualistic, would want to experience nature, would be more interested in natural and authentic experiences and less interested in mass impersonal services. The emphasis is not on more – but on experiencing greater variety (Poon, 1993).

The term millennials defines the group of adults aged 18–34 in 2015. In the US they have surpassed baby boomers in population (Fry, 2016), and constitute 20% of international tourists with spending that has increased by nearly 30% since 2007 (Fromm, 2016). Millennials travel more and often spontaneously (Weissmann, 2015), and are more interested in experiencing than in having. Adventures and opportunities to get immersed in local culture are important to them.

'New tourists' and millennials represent a sea change for hospitality. They know more and they want more, but their focus is quality rather than quantity. Many are 'digital natives' (Prensky, 2001), and they are accustomed to having access to technology and using it to not only drive their decision making, but to influence the decision making of others. In fact 97% will post to the Internet while travelling, and 75% will post once a day (Price, 2016). These individuals have a lot of power to influence the decisions of millions of potential consumers and businesses may lose patrons if they are deemed to be unsatisfactory or inauthentic. Well-known establishments may, for example, be labelled as 'tourist traps' to be avoided, and little known eateries or bars may be promoted as offering more authentic experiences.

McDonaldized experiences and staged authenticity may leave 'new tourists' longing for less-contrived encounters (Gilmore and Pine, 2007). In a world of high tech, many of these tourists will be seeking 'high touch' (Naisbitt *et al.*, 1999), and they believe that it is best achieved by interacting directly with the local community. They may not, for example, want the hotel's version of Bajan souse seasoned and served to fit European and American standards. Rather, they may want the true Bajan experience when they buy it in a local cookshop. They are aware that the cookshop may not be at the standard of their hotel, and that the exchanges between locals may be ribald or difficult to understand, but they still want to be a part of that experience.

The rise of Airbnb is a perfect example of this hunger for more authenticity and personal interaction. The front page on their website invites visitors to 'Book homes from local hosts in 191+ countries and experience a place like you live there' and to 'Belong anywhere'. Despite occasional horror stories of Airbnb experiences, the brand has grown exponentially from 47,000 people staying with an Airbnb host in the summer of 2010 to 17 million in the summer of 2015 (McAlone, 2015). Price is undoubtedly a factor, but so is the opportunity to get away from commoditized experiences. 'New tourists' will also be attracted by community visits and farm-to-table experiences where they interact with the local community. Many of these off the beaten track experiences involve locals who are not in the formal tourism sector but who are happy for the opportunity to earn from it while carrying out their daily routines.

Airbnb and similar experiences are attractive to 'new tourists' but may be a nightmare for tourism authorities who are charged with ensuring quality and safe experiences in the destination. It is nearly impossible, for example, to ensure that every street-side vendor meets even minimum safety and hygiene standards and delivers service with a smile. Similarly, it is difficult to ensure that hostels or Airbnb hosts deliver what they promise, and take the duty of care that is expected of a host. Authorities may respond by trying to ban these initiatives but realistically they cannot, because in the collaborative economy vendors and sellers can bypass them and interact directly with one another. Authorities must therefore manage these new entities in a way that facilitates the experiences that 'new tourists' want, but at the same time minimizes the risks to the destination's reputation.

'Old tourism' and 'new tourism' represent ideal types. In reality most tourists will not neatly fall into either category and this complicates the situation for tourism authorities because they must deal with a varied cross section of tastes and preferences. At the same time it represents an opportunity because it allows different types and sizes of businesses to participate in the hospitality sector. While, as discussed, 'old tourism' facilitated the growth of large hospitality businesses and branded entities, 'new tourism' may favour smaller entities that are able to provide tourists with the personalized and authentic experiences that they value. These diverse businesses must be managed in a way that adds value to the destination and to the tourists. In the following sections the key expectations for service satisfaction and the implications of these are identified. The role of the public and private sector in ensuring that these expectations are met is discussed. The chapter concludes with a discussion of the necessary actions for the future.

5.3 Creating Quality Hospitality Experiences

Hospitality businesses must create experiences that customers judge to be at a reasonable level of quality. According to the SERVQUAL GAP theory (Zeithaml, Berry and Parasuraman, 1993) reliability, assurance, tangibles, empathy and responsiveness are the elements on which customers assess service quality. The theory assumes that customers have specific expectations of each of these elements and measure their actual experiences against those expectations. The resulting gap, assuming that expectations and actual experiences are not identical, may be positive or negative.

Hospitality businesses may try to influence the expectations held by customers through advertising and by giving the customers cues via the architecture, decor, staff dress and ambiance. These are only cues, however, because expectations are so individualistic. Further, the promises made about hospitality by tourism authorities will also become the expectations about individual businesses. When Aruba calls itself the 'One Happy Island' Telegraph (2017), for example, tourists would obviously expect friendly and happy service from hospitality businesses even if they do not make this explicit promise themselves. Controlling expectations is therefore difficult and hospitality businesses will have greater control over the actual service experience and the safest course of action for businesses is to try to exceed the customers' expectations in each of the following SERVQUAL dimensions.

Reliability

The ability to deliver products and services reliably will be impacted by the design, infrastructure, equipment, system and processes. 'Old tourists' demand reliability and therefore businesses catering to them must be able to maintain equipment and upgrade when necessary. They will also require standard operating procedures (SOPs) to guide their operations. Businesses that are in a weak financial position will find it difficult to maintain their plants at the required standard and so may be forced to reduce their sales price to attract budget minded tourists. They may eventually enter the small business competitive cycle with price cutting and price wars, decline in services and products and a decreased perception of value (Crick, 2006). Reliability may be less critical to 'new tourists' since they are more spontaneous, but failures in critical aspects such as opening hours and inclusions may not be tolerated.

Assurance

The ability of employees to convey trust, their knowledge and courtesy are important to both 'new tourists' and 'old tourists' but in different ways. While 'old tourists' need to be assured that hosts know how to deliver services efficiently, 'new tourists' will be interested as well in their host's knowledge of the culture and willingness to share it. Assurance will obviously be influenced by training and the presence of SOPs and by the level of motivation to perform the jobs, as employees who are bored or disinterested may not appear assured even if they are competent.

Tangibles

Tangibles are a signal of quality for 'old tourists' and in branded and chain organizations tangibles also represent familiarity to customers through logos, consistent design, uniforms and so on (Gutek, 1995). 'New tourists' will be interested in tangibles that are authentic even if they are not in perfect condition. This means for example, that Oistins in Barbados – a popular beach restaurant setting for locals and tourists alike, may actually be valued *because* of its rustic furniture and simple setting. 'New tourists' and millennials may not be adverse to luxury, however. The upscale Island Outpost brand for example, is a collection of boutique hotels that have been described as 'a close-knit family of wildly attractive, intelligent and anarchic kids' (Islandoutpost.com). While part of a chain, each hotel is different enough to avoid the cookie cutter label disliked by 'new tourists'.

The expectations of tangibles in hospitality change frequently in keeping with technological, social and demographic trends. While certain hotel room amenities such as Wi-Fi and coffeemakers may not generate excitement, their absence is likely to become a source of annoyance and so they should be considered standard. To be competitive, hospitality businesses must also keep up with changing trends and reconfigure and refurbish regularly to accommodate them. Tangibles such as luxury linen and spa quality soaps may also be used in rooms as a source of competitive advantage. These are easily copied however so businesses must be creative to retain their competitive advantage.

Empathy and responsiveness

The provision of caring, individualized attention and the willingness to provide prompt service to customers requires having well trained and caring staff who are attentive to customer needs. Gutek (1995) argues that there are three types of interactions – *encounters* where the service employee and the customer are strangers and have one off interactions; *relationships*, which are of a continuing duration and where there is a history of shared interaction and knowledge of one another; and *pseudo-relationships*. In pseudo-relationships there is no history between the service providers and the customers but staff attempt to make customers feel as if they are in a relationship. Hotels like the Ritz Carlton, for example, use technology as an aid in storing guest information that is then used to provide customized service.

While encounters and pseudo-relationships may be appealing to 'old tourists', they may be less so to 'new tourists' who prefer responses based on genuine relationships. Relationships, while rich in interaction, are a costly investment in time and emotions and may be somewhat difficult for managers to control. Hospitality businesses have to strike the balance between their need for control and cost effectiveness with some customers' preferences for deeper interactions.

Most hospitality organizations are privately owned and operated and therefore personally responsible for ensuring that these standards are met. The destination may suffer if customers are not satisfied and therefore the public sector needs to play a role in ensuring quality standards. We will now explore how this role may be carried out.

5.4 Role of the Public and Private Sectors in Ensuring Quality Products and Services in Hospitality

Recruitment, selection and retention

While hospitality has a reputation for low-skilled employment (Guerrier, 1999), the requirements of even entry-level jobs are increasing as room attendants and waiters may have to use hand-held tablets to communicate, for example. Further, even back of the house employees are expected to become part of the 'show' and are expected to engage customers they encounter. Soft skills are therefore important in every area. With increasing diversity in tourists, hospitality workers may also need to learn additional languages and to be sensitized to the customs and expectations of different cultures and groups.

A 1999 study of tourism training needs in the Caribbean identified soft skills including customer relations and communication as the most important training need for all groups except managers (Charles, 2002). Technical or hard skills are also critical if the Caribbean is to compete effectively in tourism, and hospitality workers will also need to be productive and efficient. Finding the combination of all of these skills can be difficult in many small Caribbean destinations and for many hospitality businesses the challenge has often been to attract the brightest and the best to work for them. This problem is exacerbated by working conditions such as long and non-family friendly working hours, challenging work requirements and sometimes low pay as well as a lingering resentment about the industry (Patullo, 1996). That explains why parents and teachers sometimes discourage young people from entering the industry (Crick, 2008). A final hurdle that employers must overcome is the perception that there is a glass ceiling for locals. This perception, harkening back to the early days of tourism when many or most of the top positions were filled by expatriates, needs to be corrected if talented entrants are to view the industry as a lucrative career choice.

Caribbean governments and hospitality managers have recognized the challenges of attracting the right people to the sector and have developed a number of interventions in response. The Caribbean Tourism Organization (CTO) has developed a manual on hospitality career opportunities and the Bahamas has publicized more than 1000 careers in tourism[2] with many of those careers being in hospitality. Scholarships for tertiary education are offered by public and private sector entities and around the region hotel and tourism associations work closely with schools to educate young people about work in the industry. Some hotels also define themselves as 'Employers of Choice' in order to be more appealing to potential applicants (Crick, 2008).

Initiatives such as these increase the pool of available talent for the sector but selecting the right people will depend on effective screening mechanisms. Psychometric testing may be an option for those entities that can afford them. Entities with more limited resources may, however, find employees by tapping into their network by utilizing family connections or recommendations by existing employees (Crick, 2008). Governments may assist businesses of all sizes by creating a pool of willing and able potential employees. Barbados' National Initiative for Service Excellence (NISE), for example, uses different methods to sensitize *all* Bajans to world-class service standards. As a result, the entire population is cognizant of the level of service expected and this increases the likelihood that better service will be provided. Jamaica's Tourism Infusion programme and Tourism Action Clubs take a different approach by introducing students to the benefits of tourism, and by helping them to see tourism jobs as a viable option.

Turnover, generally a problem in hospitality, is generally less so in the Caribbean where high rates of unemployment and limited options, especially for those with minimum education and training, reduce mobility. Reduced mobility is a mixed blessing, however, if businesses only retain those that cannot find employment elsewhere. Hospitality managers must therefore find creative ways of retaining competent employees. Larger organizations are more likely to have structured motivational programmes in place than smaller ones, but as Crick (2008) noted, managers of smaller properties may be able to engender loyalty from employees who feel a sense of family and belonging and are empowered in their jobs. If hospitality is to be viewed as a career rather than a job of last resort, hospitality businesses of all sizes will need to develop career ladders and attractive pay and benefits. Hospitality businesses may of course fill the gaps by recruiting non-nationals. This may add diversity of skills, knowledge and culture to enterprises but must be carried out in compliance with local labour laws and with appropriate care taken to avoid alienating local job seekers.

Highly skilled employees, particularly chefs, bartenders and some restaurant personnel, may be especially targeted by other local employers and increasingly by overseas agents who work for cruise ships and overseas hospitality organizations. These recruiters have been viewed as a problem, but this 'problem' may present an opportunity for the Caribbean, and governments may wish to explore increasing the training capacity of local institutions to create an oversupply of those skilled persons that are particularly targeted.

Training and development

Hospitality is a dynamic sector and therefore continuous training and development will be required if businesses are to remain current. Caribbean governments already play an important role in supporting the development of hospitality institutes and training hotels such as Barbados' Pommarine Hotel and Bermuda's Hospitality Institute. These institutes and national training agencies help in establishing minimum standards of performance and training in best practices. In-house training will, however, always be necessary and businesses, particularly smaller ones, may have to be supported in developing the requisite training skills. While some businesses may be reluctant to invest in training employees because they fear that they will be poached by other businesses, they should be discouraged from this short-sighted view that can erode quality and harm the destination's reputation for quality products and service.

Financing

Setting up and maintaining hospitality businesses can be expensive, particularly since most of the equipment and sometimes much of the stock has to be imported. Access to affordable business financing will enable greater participation of businesses of all sizes in the sector and will also allow businesses to retool and remain competitive. Financial entities will, however, need to be sensitized to aspects of the business-like seasonality. They may also have to be encouraged to provide assistance to the smaller start-ups. In this regard non-traditional financing such as crowdfunding and angel investors may be a useful complement to the traditional financial sector. Governments may also play a role by providing tax relief for upgrading equipment and renovating the physical plant.

Legislation and regulation

Hosts have a significant legal duty to safeguard the welfare of their guests. Beyond that, however, they may face significant financial and reputational penalties if they fail to meet expectations or put their guests in danger. Governments play a critical role by sensitizing hospitality operators to risk management and by establishing minimum standards that must be adhered to. At the same time, however, governments must exercise good judgment in setting and enforcing regulations. While the local rum shop for example, must meet basic hygiene and safety standards, it should not be expected to have all the facilities of a large sports bar. Indeed if it has all of these facilities it may lose the authenticity that makes it attractive to 'new tourists'.

Partnerships

Hospitality businesses may choose to compete based on differentiation, focus or low cost (Porter, 1998). Competition based on low cost may not be amenable to partnerships and alliances, but competition based on differentiation and focus may be. As the diversity in the market increases it is possible for hospitality businesses with a focus and differentiation strategy to collaborate to ensure that visitors have an all-around positive experience. Bay Gardens hotel in St. Lucia has, for example, differentiated itself by partnering with restaurants in the surrounding community to create a 'dine-around' experience for all-inclusive and guests on a meal plan. This provides their guests with more options and allows local restaurants to play a greater role in tourism. In Jamaica, all-inclusives facilitate access to some community experiences where their guests may take a day trip to a community where they eat Jamaican food, meet with local farmers and producers and have a cultural experience. These partnerships will be particularly important to 'new tourists' who are less likely to be satisfied with the offerings of a single entity. Other partnerships may involve training and joint marketing, and even mentorship of new business operators by more experienced ones.

5.5 Lessons Learned

The evolving Caribbean hospitality sector presents opportunities and challenges. The diversity in the market means that there is scope for hospitality businesses of different sizes and offering a range of products and services. The Internet also levels the playing field making publicity relatively easy and cheap for small entities. There are challenges ahead, however. The Internet also makes it easy for dissatisfied visitors to share their complaints to a wide market and this means that there is little room for error. Businesses, especially the smaller ones, may need government support in putting in place systems and procedures to reduce risks and ensure consistency. Businesses may also need to be supported in gaining access to the financial resources that will allow them to meet international standards and to expand if they need to do so.

People are at the heart of hospitality and much of Caribbean tourism and hospitality is predicated on their warmth and goodwill. The lines between tourists and hosts are being blurred as tourists deliberately seek out new experiences and interaction with locals. Airbnb, for example, will significantly increase the number of tourists who are staying in private homes, and trends such as 'farm to table' and buying local will also mean that many tourists are straying away from traditional tourist areas into the heart of the country.

The goodwill of *all* locals will therefore be critical. While tourism is essential to most Caribbean islands, the citizens of the destination have not always felt that they benefit from the industry and are also sometimes unhappy about the strategic direction of tourism. Tourism authorities will need to address these concerns because if they do not then they will lose goodwill and support. While 'old tourists' could be shepherded to where they were assured of a favourable welcome, 'new tourists' are curious and want to go off the beaten track and to meet locals who are not part of a staged experience. When they do, it is important that they have a favourable experience.

In the coming years Caribbean hospitality will require an innovative and collaborative response from hospitality businesses and governments. The good news is that the emergence of 'new tourists' and millennials has expanded the pie to include many more hospitality operators. There are greater opportunities for the many players to work collaboratively to create a vacation experience that is safe, interesting and appealing to the increasingly diverse hospitality market. Governments will need to play a role in brokering alliances that will be mutually beneficial and in promoting and supporting micro and small hospitality businesses so that they play a greater and more profitable role in tourism.

Notes

[1] McDonaldization was a term coined by George Ritzer to describe the influence and inexorable force of the fast food chain McDonalds. It has four tenets - efficiency, calculability, predictability and control. Examples may be seen in restaurants, chain outlets, schools and even medicine.

[2] The Bahamas Hotel Association has compiled a list of career opportunities within the tourism industry, which can be accessed at www.bhahotels.com/public/tourism-careers.pdf

References

Charles, K. (2002) Future human resource development needs of the Caribbean tourism industry. In: Jaywawardena, C. (ed.) *Tourism and Hospitality Education and Training in the Caribbean*. University of the West Indies Press, Kingston, Jamaica, pp. 139–152.

Crick, A. (2006) EP/Small Hotels: Prospects for the future. In: Hall, K. and Holding, R. (eds) *Tourism: The Driver of Change in the Jamaican Economy?* Ian Randle Publishers, Kingston, Jamaica, pp. 71–91.

Crick, A. (2008) No plantation work here: contemporary HR practices in Caribbean hotels. *International Journal of Contemporary Hospitality Management* 20(1), pp. 79–89.

Croes, R. (2006) A paradigm shift to a new strategy for small island economies: Embracing demand side economics for value enhancement and long term economic stability. *Tourism Management* 27(3), pp. 453–466.

Croes, R. (2011) *The Small Island Paradox: Tourism Specialization as a Potential Solution*. Lap Lambert Academic Publishing, Saarbrücken, Germany.

Fromm, J. (2016) *Luxury Travel Trends Shift as Affluent Millennials Act on Their Aspirations.* [online] Available at: http://www.forbes.com/sites/jefffromm/2016/03/29/luxury-travel-trends-shift-as-affluent-millennials-act-on-their-aspirations/#3a22074c22b4 (accessed 24 May 2016)

Fry, R. (2016) *Millennials overtake Baby Boomers as America's largest generation.* [online] Available at: http://www.pewresearch.org/fact-tank/2016/04/25/millennials-overtake-baby-boomers/ (accessed 24 May 2016).

Gilmore, J.H. and Pine, B.J. (2007) *Authenticity: What Consumers Really Want*. Harvard Business Review Press, Boston, Massachusetts.

Guerrier, Y. (1999) *Organizational Behaviour in Hotels and Restaurants. An International Perspective.* Wiley, Chichester, UK.

Gutek, B. (1995) *The Dynamics of Service: Reflections on the Changing Nature of Customer/Provider Interactions*. Jossey-Bass, California.

Jacob, D. (2011) Mario Sabga-Aboud: the man who's made rituals a way of life. [online] Available at: http://caribbean-beat.com/issue-109/man-whos-made-rituals-way-life#axzz4Ui9I6ZWh (accessed 2 January 2017).

King, B. (1997) *Creating Island Resorts*. Routledge, New York.

Lashley, C. (2001) Towards a theoretical understanding. In: Lashley, C. and Morrison, C. *In Search of Hospitality: Theoretical Perspectives and Debates*. Butterworth-Heinemann, Oxford, UK, pp. 1–17.

McAlone, N. (2015) *This chart shows exactly how insane Airbnb's growth has been over the past 5 years* [online] Available at: http://www.businessinsider.com/airbnbs-summer-reach-has-grown-by-353-times-in-5-years-2015-9 (accessed 16 June 2016).

McElroy, J. (2006) Small island tourist economies across the life cycle. *Asia Pacific Viewpoint* 47(1), pp. 61–77.

Naisbitt, J., Naisbitt, N. and Phillips, D. (1999) *High Tech, High Touch: Technology and Our Search for Meaning*. Broadway Books, New York.

Patullo, P. (1996) *Last Resorts: The Cost of Tourism in the Caribbean*. Ian Randle Publishers, Kingston, Jamaica.

Pine, B. and Gilmore, J. (1999) *The Experience Economy: Work is Theatre and Every Business a Stage.* Harvard Business School Press, Boston, Massachusetts.

Poon, A. (1993) *Tourism, Technology and Competitive Strategy*. CAB International, Wallingford, UK.

Porter, M. (1998) *Competitive Strategy: Techniques for Analyzing Industries and Competitors*. The Free Press, New York.

Prensky, M. (2001) Digital natives, digital immigrants. *On the Horizon* 9(5), pp 1–6.

Price, W. (2016) *Millennial Travel Trends: A Look at the Largest Generations' Habits.* Available at: http://www.internetmarketinginc.com/blog/millennial-travel-trends/ (accessed 15 June 2016).

Ritzer, G. (1996) *The McDonaldization of Society: An Investigation into the Changing Character of Contemporary Social Life*. Pine Forge Press, California.

Sandals (2017) Sandals The Only All-Inclusive to Include it All. Available at: http://www.sandals.so.uk/home (accessed 31 October 2017).

Telegraph (2017). One Happy Island. Available at: http://www.telegraph.co.uk/travel/destinations/caribbean/one-happy-island/ (accessed 31 October 2017).

Weissmann, A. (2015). Today's (and Tomorrow's) Millennials. Available at: June 15, 2016, from travel-weekly.com: http://www.travelweekly.com/Arnie-Weissmann/Todays-and-tomorrows-millennials (accessed 15 June 2016).

Zeithaml, V., Berry, L. and Parasuraman, A. (1993) The nature and determinants of customer service. *Journal of the Academy of Marketing Science* 21(1), pp.1–12.

PART II: Island Tourism Policy, Planning and Development

6 A Comparative Analysis of Tourism Policy Networks

Michelle McLeod,[1]* Donna Chambers[2] and David Airey[3]

[1]*The University of the West Indies, Jamaica;* [2]*University of Sunderland, United Kingdom;* [3]*University of Surrey, United Kingdom*

6.1 Introduction

Policy making in tourism is the process of formulating tourism policies that will guide tourism growth and development. The policy making process is often viewed as a consultative one that engages a number of stakeholders and can thus be perceived as a policy network (Dredge, 2006; Pforr, 2006). An output of the policy making process is tourism policy content that forms part of public policy that Hall and Jenkins (2004) describe as involving government action as a political activity. Tourism policy content is defined as regulations, guidelines, directives, objectives and strategies to affect tourism development (Ritchie and Crouch, 2003). Based on policy guidelines derived from consultative processes with stakeholders, mandates are given to certain implementation agencies to fulfil the intentions of tourism policy.

Stevenson, Airey and Miller (2008) note that tourism policy making is a social process that involves communication, negotiation and consensus building. Consensus is needed to move forward for the implementation of effective tourism policies. The divergent views about which policy directions take mean that policy making as an activity requires attention. Yet, there has been limited examination of the policy making process itself as to whether the tourism policy content is appropriate to affect tourism growth and development.

A network perspective can therefore add value to understanding the tourism policy making process. Several researchers have examined these networks (Dredge, 2006; Pforr, 2006; Dredge and Pforr, 2008; Wong, Mistilis and Dwyer, 2010; Dela Santa, 2013). The interrelationships of stakeholders in the tourism sector occur in several different ways. First, there are formal links through the government and non-government agencies including the private sector. Second, there are interlocking directorships, which may not be designed as pre-determined formal arrangements but often become formal in

*E-mail: michelle.mcleod@uwimona.edu.jm

 Tourism Management in Warm-water Island Destinations (eds M. McLeod and R. Croes)

nature since directors may have been officially appointed to the government agencies to serve in a particular capacity on state or non-state boards. Third, there are informal links resulting from the personal relationships of tourism stakeholders through for example political affiliation, alumni links, associations, groups and societies.

While acknowledging the importance of informal links, it is the formal relationships that are the focus of attention here. This is because formal relationships are generally forged by the purposeful design of the policy actors in order to implement, in this instance, tourism policy. Further, formal structures are more likely to provide a basis for interactions around the policy making agenda than informal structures. Indeed, although formal and informal networks exist within tourism destinations (Strobl and Peters, 2013; Pforr, Pechlaner, Volgger and Thompson, 2014), formal network structures around an organization or board of directors form higher network densities (Pforr, 2006). Also, Strobl and Peters (2013) considered that while informal relationships may be safe based on trustworthiness, formal relationships based on their legal character are safer. In the context of tourism, which comprises so many different stakeholders, formal relationships are of prime significance in policy making and have a particular time-frame for existence that can allow for an examination of the results that emerge from these relationships. This is because such relationships normally exist over a longer period of time, whereas informal relationships are more fickle in terms of establishing policy making influences. Against this background, this chapter seeks to examine the formal network relationships that influence the tourism policy making process. It commences with a brief review of the literature on social network analysis, which provides the foundation for exploration of policy networks, before proceeding to a discussion of tourism policy and policy making. In order to illustrate the complexities of the tourism policy making process and the importance of networks within this context, the next section of the chapter provides an empirical analysis of formal tourism policy networks in two Caribbean countries – Jamaica and Trinidad and Tobago. The chapter concludes with a summary of the key findings and the central lessons that have been learnt about the influence of networks on tourism policy making.

6.2 Literature Review

Social network analysis

Based on the network perspective, the analysis is broadly underpinned by the theory and methodology of social network and its analytical techniques. A social network refers to the relationships among social entities or actors. Social network analysis (SNA) thus seeks to understand the nature, patterns and implications of these relationships (Wasserman and Galaskiewicz, 1994; Otte and Rousseau, 2002). According to Scott (2013), SNA first developed in the mid-20th century from the structural concerns of eminent English social anthropologist Alfred Radcliffe-Brown. Mizruchi (1994) highlights its interdisciplinarity through having roots in several theoretical perspectives in psychiatry, anthropology, structural sociology and the structuralism of Levi-Strauss. Scott (2013) also points to the recent link between SNA and the theory of social capital. In this sense, social networks are a 'particular form of social capital that individuals can employ to enhance their advantages and opportunities' (Scott, 2013, p. 8). However, he argues that social networks are more than that as they may be also networks of economic transactions and political conflicts. For Scott, 'Social network analysis must be seen as a comprehensive and all-encompassing approach to the relational features of social structures.' (2013, p. 8).

In a similar vein, Wasserman and Galaskiewicz (1994) argue that SNA does not focus on analysing individual behaviours, attitudes and beliefs but on the relationships among actors in interaction with each other and how these constitute a structure that can be studied in its own right. They suggest further that this kind of analysis makes certain assumptions: first, that actors and actions are interdependent and not autonomous units; second, that relational ties among actors act as channels for the flow of resources; and third, that structures form enduring patterns of relationships among actors. Knoke and Yang (2008) elaborate on these assumptions by claiming that structural relationships are often more significant for understanding observed behaviours than are attributes such as values and ideologies. Further, they indicate that perceptions, beliefs and actions are affected by these social networks through various structural mechanisms that are socially constructed through the relations among entities; and structural relations should be seen as dynamic processes. Social networks exist at the micro level as in egocentric or personal networks and at the macro level of organizations, institutions or even whole countries.

It is evident that SNA has relevance for a plethora of contexts of social life, including the political. Indeed, social and political reality is complex and strongly interconnected (Kenis and Schneider, 1991). In this sense SNA can be adapted to illuminate the nature of the policy making process, and importantly the concept of policy networks became popularized at the end of the 1970s as a result of significant changes in the political governance of modern democracies (Kenis and Schneider, 1991). As early as 1978, Katzenstein envisaged a policy network as akin to:

> a political metastructure integrating different forms of interest intermediation and governance, forming a symbiotic relationship between state and society in policy making (Kenis and Schneider, 1991, p. 31).

Much later, Rhodes, in a similar vein, suggests that a policy network can be viewed as:

> ...formal institutional and informal linkages between governmental and other actors structured around shared if endlessly negotiated beliefs and interests in public policy making and implementation (2006, p. 426).

In tourism, as noted later in this chapter, SNA has also been used to understand the relationships among actors within tourism policy making.

Tourism policy and policy making

Dye's often quoted definition of policy as being 'whatever governments choose to do or not to do' originally articulated in 1972, but subsequently appearing in later editions of his text (2008, p. 3), provides a clear description. However, at the same time it misses a key and increasingly important aspect of public policy that has been recognized through the use of a network approach, i.e. that it is not government alone that has a monopoly on public policy. As Airey and Ruhanen suggest in relation to tourism, it misses the point that policy involves 'complex interacting elements' (2014, p. 150). Policy eventually is the outcome of negotiations between stakeholders including, but not confined solely to, government. These complex interacting elements provide the context for this research.

Tourism policy content is an important consideration to effect tourism development. Jenkins (1991) notes that tourism policy can be distinguished based on different policy directions for tourism development. He created a framework which suggests that these directions can be: public or private sector driven; focused on international or domestic tourism; and/or lead to integrated or enclave tourism. The nature of the direction taken within a particular tourism policy framework is important since this will contribute to the design and involvement of implementation agencies. For instance, if sustainable tourism is an overarching tourism policy direction, then agencies relating to climate change and environmental management practices may emerge to manage its implementation.

The process of changing government policy seems to be ad-hoc and occurs when a problem or opportunity arises. Fayos-Solá (1996) calls for a balanced partnership between stakeholders in the policy making process. To move beyond a 'midsummer or ephemeral perception of a tourism policy', Fayos-Solá (1996, p. 409) suggests broad-based input from a range of tourism actors. Public sector intervention alone is not adequate and several authors indicate that networks often emerge to formulate tourism policy (Tyler and Dinan, 2001; Pforr, 2006). Tyler and Dinan (2001) note that a tri-axial network comprising three sub-networks operated as a policy network in England including government, public resource management and commercial tourism groups. These sub-networks were interconnected with links between government and public resource management and, government and commercial tourism, with a weak link between public resource management and commercial tourism (Tyler and Dinan, 2001); relationships between network agents are described as immature and thus perhaps emerging. Pforr (2006) studied a sub-set of 54 tourism sector organizations to understand their connections within the context of tourism master planning. Several exchange relationships emerged as a result of the connections between these organizations. These connections determined the planning processes of the public, private and not for profit agents. The study concludes that policy making was influenced by political interest and tourism sector priorities.

In summary, the literature suggests that SNA can be adapted to analyse the policy making process in tourism and can reveal the complex range of networks in tourism, especially those underpinned by formal relationships. The literature also illustrates the importance of network structures thus establishing the theoretical context for this study. Within small island destinations policy networks are particularly important to understand as an individual can play multiple roles and be a member of several organizations at the same time. This type of formal interconnection in small island destinations can at the same time provide an opportunity to build consensus and also constrain the provision of new ideas in the policy making process. Two Caribbean countries, Jamaica and Trinidad and Tobago, provide contrasting settings to undertake this exploration of the complex interrelationships involved in tourism policy making. The chapter now turns to the methods before reporting on, and discussing the findings.

6.3 Tourism Policy Networks Methods

Research design

This chapter explores the agencies involved in tourism policy making in two Caribbean countries using SNA. In so doing, the chapter also illustrates the importance of tourism policy formulation and the goals achieved in terms of decision making. Networks involve stakeholders in the development of tourism policy and these stakeholders have access to tourism information that is utilized for tourism policy making. As discussed, SNA was utilized since this approach broadly examines the interconnections between tourism policy actors in tourism policy making. UCINET software was utilized for analysis (Borgatti, Everett and Freeman, 2002) and NetDraw software for illustration (Borgatti, 2002).

There are two social network research designs: a whole network design; and an ego or personal network design (Borgatti, Everett and Johnson, 2013). The latter exists at the micro level and consists of a focal node (ego) and the nodes with which it is directly connected (alters) along with the ties, if there are any, among the alters. The former exists at the macro level and the focus here is on the design of entire networks and the recognition of structural positions and components of the network (Wasserman and Galaskiewicz, 1994). Given the preoccupation of the current study with tourism policy networks among formal institutional actors, it is apposite to draw on a whole network approach. This approach enables the inclusion of comprehensive connections between the policy actors (nodes), to explore their involvement with implementing tourism policy and reduces the exclusion of actors and ties that can influence the policy making process.

Agencies were categorized using Hall's (2011, 2012) elaboration of a governance typology comprised of: hierarchies in the public sector; markets in the private sector; networks of partnerships; and communities and citizens groups. Hierarchical governance focuses on the legislative and regulatory role of the state (and supranational institutions) in a top-down process of governance. This is seen as a traditional approach, which does not adequately account for the role and power of non-state actors (largely the private sector) in contemporary societies. Indeed, the rise and ubiquity of neo-liberalism has led to the privatization of much of the tourism industry and tourism functions and thus the market has emerged as a key player in tourism governance. Networks refer to a more collaborative form of governance where partnerships are created between public and private sector stakeholders. Network governance is seen as a sort of 'middle ground' between hierarchies and markets, which can lead to a more integrated approach to tourism policy making. Community governance exists where the involvement of local citizens within tourism policy making and implementation is greater and more direct. This is seen as essential for more sustainable forms of tourism development (see Hall, 2011, 2012). To provide an additional context, in this chapter distinctions were made between international, regional and local agencies. The agencies' influence were mapped and in the mapping exercise several formal relationships that could influence tourism policy making emerged in terms of reporting, inter-board and information sharing relationships.

Case study areas

The tourism industry exhibits different characteristics and patterns between the two case study areas and also between the twin islands of Trinidad and Tobago. Jamaica is a mature tourist destination and the industry there is said to have blossomed from the late 19th and early 20th centuries. This was due to the concerted efforts made by Government to promote the industry and by private investors to establish large hotels. While the accommodation sector is largely Jamaican owned, in recent years the industry has witnessed a trend in the construction of large hotels by Spanish hotel chains (Planning Institute of Jamaica, 2009). All-inclusives are the main type of accommodation also accounting for the majority of room capacity and significantly higher rates of room occupancy. Considerable progress has been made in achieving growth targets but other areas such as community-based development and environmental sustainability have not witnessed commensurate progress (Planning Institute of Jamaica, 2009).

Trinidad and Tobago as a twin island state has mixed stages of tourism development. Trinidad, the larger of the two islands, has an oil-based economy while Tobago is heavily dependent on tourism. In Tobago tourism accounted for 98.4% of direct exports and 47.6% of jobs in 2009 (Ministry of Tourism, 2010). Both islands also have different potentials for tourism development with Trinidad centring on its cultural diversity and business tourism, while Tobago is promoted as a 'semi-rustic, idyllic island' with a focus on the leisure tourism market (Ministry of Tourism, 2010). While there are similarities in the types of accommodation, generally Tobago has smaller, independent, non-brand properties while in Trinidad there are some large brand hotels. While recognizing the differences between both islands in terms of tourism development trajectories, the twin island republic nevertheless saw itself in 2009 as an 'emerging destination' (Ministry of Tourism, 2010) with tourism seen as having critical potential for the success and advancement of the nation.

Figure 6.1 illustrates contrasting tourism growth between the two case study countries with Jamaica on an upward growth path while Trinidad and Tobago's tourism growth is somewhat stalled on an undulating curve. Jamaica celebrated receiving over two million stop-over visitors in December 2013 while Trinidad and Tobago approached half a million stop-over visitors annually. According to the Caribbean Tourism Organization's (CTO) country statistics, Jamaica received 2.1 million stop-over visitors in 2014 and 1.4 million cruise visitors (CTO, 2015). Jamaica's main market is the United States of America with 62.3% of visitors coming from there in 2014 (CTO, 2015). Visitor expenditure was US$2 billion with a budget of US$38.6 million being spent on the sector and the average room occupancy rate was 60.5% in 2010 (CTO, 2014a).

Trinidad and Tobago's Carnival is a major generator of visitors and expenditure. In 2012, over the Carnival period from February 3rd to 21st, 38,252 visitors were recorded (CSO, 2014). The country received 412,537 visitors in 2014 with the United States of America accounting for 38.4% of those visitors (CTO, 2015). Cruise passenger arrivals were just over 42,820 excursionists (CTO, 2015). The CTO notes a tourism budget of US$4.1 million and a tourist expenditure of US$366.6 million (CTO, 2014a). This tourist expenditure is 18.3% of the tourist receipts of Jamaica. The average room occupancy was 52.0% in 2010 (CTO, 2014a).

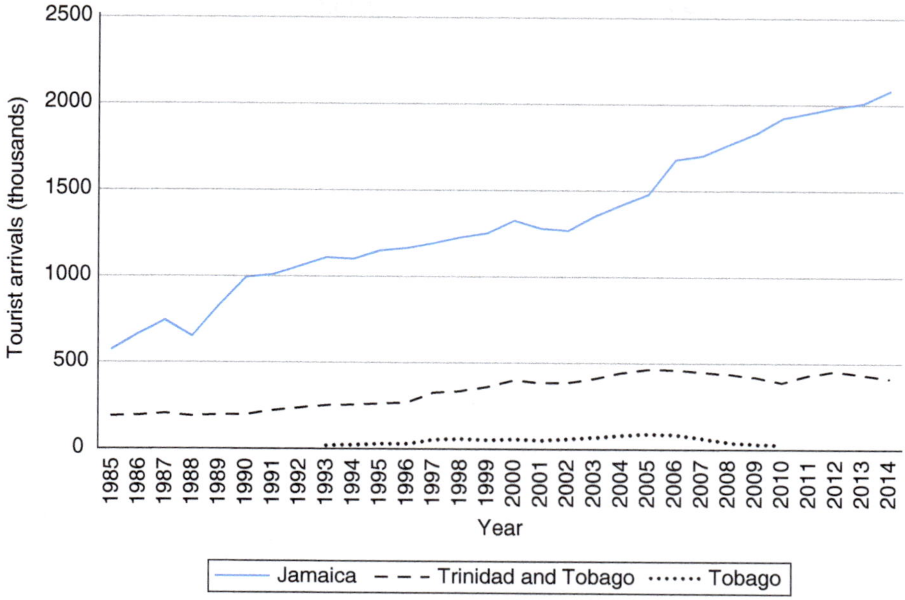

Fig. 6.1. Stop-over visitor arrivals to Jamaica and Trinidad and Tobago (1985 to 2014). (From Jamaica Tourist Board; Trinidad and Tobago Ministry of Tourism and Trinidad and Tobago Tourism Development Company. NB Separate figures for Tobago shown for emphasis only.)

The national tourism policy of both countries can be compared to explore important differences in tourism development within these two island states. The Government of Jamaica has articulated a vision for the tourism sector that is based on the National Tourism Policy: 'An inclusive, world-class, distinctly Jamaican Tourism Sector that is a major contributor to socio-economic and cultural development, with a well-educated, highly skilled and motivated workforce at all levels within a safe, secure and sustainably managed environment' (PIOJ, 2009, p. 48). According to the Approved National Tourism Policy of Trinidad and Tobago: 'The Government of Trinidad and Tobago shall create an environment that facilitates the country's tourism growth by addressing human resource development, community development, infrastructure and transportation development, investment promotion, accommodation issues, product development, and marketing' (Ministry of Tourism, 2010, p. 27). Both policy documents have highlighted the specific policy issues that require attention for the development of tourism in the countries. Whereas Jamaica has concerns for its workforce, specific mention is made of safety and security, and for Trinidad and Tobago, the policy direction is based on a list of destination management issues to be addressed.

Data collection and analysis for tourism policy making

Data were gathered from archival sources of Caribbean tourism policy documents. Two documents formed the starting point for data collection: National Tourism

Policy of Trinidad and Tobago (2010); and Vision 2030 Jamaica Final Draft Tourism Sector Plan (2009). Secondary data sources can provide valuable information without making primary contact. Information regarding actors within a network can be obtained within a short time frame using secondary data. Another advantage is that there is no refusal or non-participation since a comprehensive list of actors is sourced. These archival documents were provided by the Caribbean Tourism Organization's library and formed the basis for identifying the agencies involved with tourism policy, planning, development and management. Once the agencies were identified and coded, the next step was to identify the interconnections among them. Interconnections were found through identification of the composition of the boards of directors and where possible an online search for the agency or agencies to which an individual director belonged. Once a director had been identified in one tourism agency with a relationship with another tourism agency, then a link was recorded between agencies.

Connections were also formed through formal arrangements such as the provision of data and information from one agency to the other and reporting relationships. Information on these came from an online search that included a list of agencies related to a particular ministry. Joint marketing and product development relationships were considered as in the case where the Jamaica product development agency (Tourism Product Development Company) works with the marketing agency (Jamaica Tourist Board) and both agencies report to the Ministry of Tourism in Jamaica. LinkedIn was a good source of information for confirming the directors of agency boards. In terms of information relationships, the Jamaica Information Service's (JIS) website also provided the links between the government bodies and their agencies. Recent newspaper articles in both countries were examined to determine any changes in the relationships of the various tourism agencies. This was particularly relevant in Trinidad and Tobago as a newspaper report revealed that there was a change in the governing political party during the period of the research. A subsequent online search revealed that one ministry closed and this information was noted.

In order to provide an external dimension to the information secured from the secondary sources, a summary of the results was presented to tourism industry officials at the 3rd University of the West Indies (UWI) International Tourism Conference, November 9–11, 2014. This provided an opportunity to refine the information obtained and for external validation of the data sources. This conference presentation highlighted the network diagrams of the Jamaica and Trinidad and Tobago tourism policy networks. It was suggested by the industry officials present that the Jamaica tourism policy network include those stakeholders that are directly responsible for tourism policy. This feedback was useful to refine the Jamaica network.

It is important to note that the setting of a network boundary must be theoretically relevant to the subject under study. While institutional stakeholder mapping of the tourism agencies in Jamaica and Trinidad and Tobago have been mapped (see McLeod, 2015), there was a need to strengthen understanding of the policy making framework in these two Caribbean countries. In this case, the tourism policy making actors and relationships between them were the relevant criteria and therefore certain actors were either included or excluded for a number of reasons. The data included those organizations primarily engaged in the business of tourism (transport, hotels and attractions), and were cleaned to remove those that were not (Appendices I and II). The Vision 2030 Jamaica Final Draft Tourism Sector Plan (2009) mentions a number of government agencies such as the Cabinet Office and Ministry of Education; however, given that these agencies were not directly relevant for tourism policy making, they were excluded from the analysis. In the case of Trinidad and Tobago, all the agencies mentioned in the source document (Ministry of Tourism, 2010) were included. This meant that certain industry associations and several tourism-related agencies that were not listed in this source document, for example, the National Carnival Commission were excluded from the analysis. In this regard, an agency is treated as non-relevant for tourism policy making if it has not been listed by the tourism agency that authored the source document utilized in this research study.

The agencies were coded by country or region, then by governance type (hierarchies, markets, networks or communities) and finally a unique identifier in the form of a number was placed at the end of the node label. The unique identifier counts the number of agencies in each country or region. Each actor can potentially have one of three different types of relationships (official reporting, inter-board and information-sharing). Each agency relationship was developed as a linked list. The data were symmetrized, which means that flows occur between two agency actors, which are connected.

Data collected about institutional stakeholder mapping have been included to strengthen understanding of the policy making framework (see McLeod, 2015). Institutional stakeholder mapping involved identification of the policy issues in the tourism sector and the stakeholders involved in its development and management. Connections were then made between the policy issues in the Caribbean region, capacity management, economic linkages, environment, health and safety, marketing and transportation (Caribbean Tourism Organization, 2014b), and the actors involved with addressing these issues. The final step in the institutional stakeholder mapping involved validation of stakeholder maps by interviewees. This draws upon an earlier study (McLeod, 2015; McLeod and McNaughton, 2016) that shows these stakeholder maps in the tourism sector for five Caribbean countries including Jamaica and Trinidad and Tobago.

6.4 Policy Making in Two Caribbean Countries

This work furthers understanding of the tourism policy making interactions within the tourism sector of the countries as the data show the actors involved include three types of hierarchical relationships among agencies: an official reporting relationship; an inter-board relationship; and an information sharing relationship by virtue of being in the same region, industry or practice. Figures 6.2A and 6.2B show the tourism policy networks diagrammatically, for both fieldwork sites. For each they provide the numbers of actors, their composition and structure. The findings show that there were important differences between the two countries. These figures show that the Jamaica tourism policy network includes 34 actors, with 27 actors in Trinidad and Tobago and that the shape of the structures vary with Jamaica's tourism policy network being more circular (Fig. 6.2A).

The cut-points (points at which an actor's removal from the network will fragment the network) have a different effect in that the consensus building required for tourism policy making and implementation can be fragmented if one particular agency that is a cut-point is removed from the policy network. Figures 6.2A and 6.2B show the blocks (blue circle) and cut-points (red square nodes) that can either combine the network structure or fragment it. In the case of a block of policy actors that are classified based on the same country or region, and governance type, any particular combined block can be assessed for the contribution of these policy actors on policy making. For example, there is an evident block of international actors in the Jamaica tourism policy network (Fig. 6.2A, bottom-left block of actors that includes

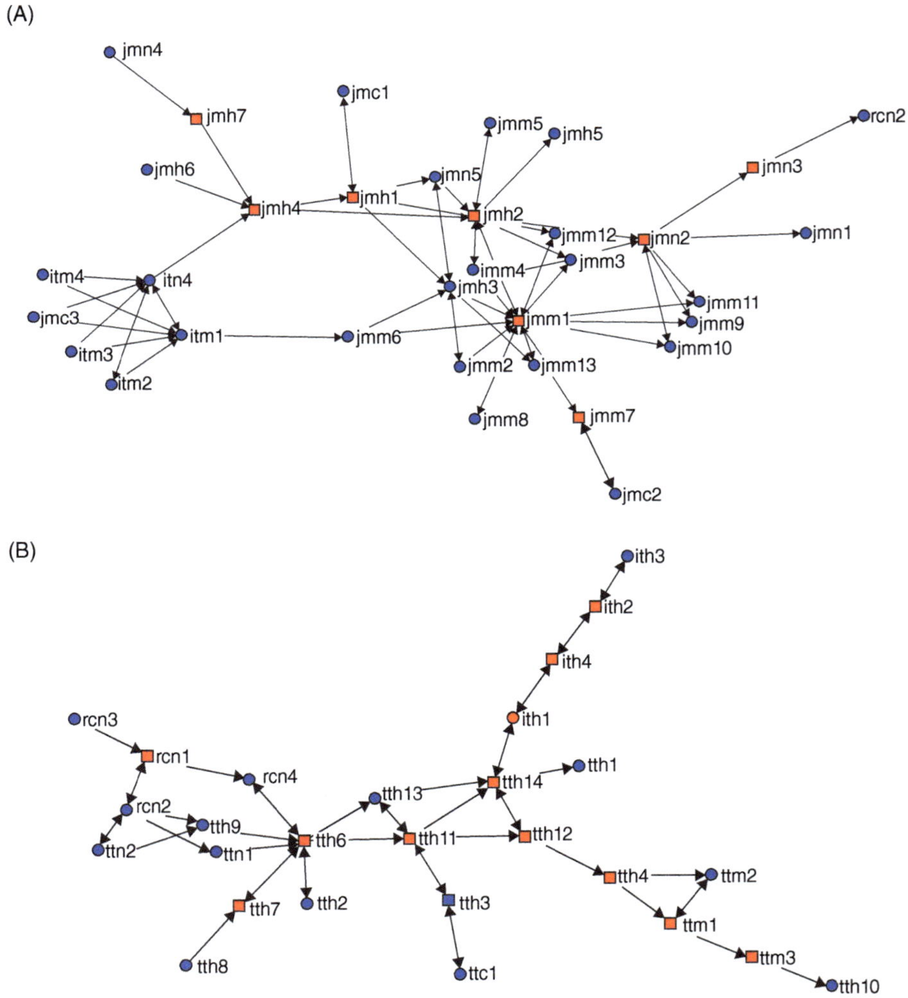

Fig. 6.2. (A) Jamaica tourism policy network. (B) Trinidad and Tobago tourism policy network. Key: Jamaica (jm); Trinidad and Tobago (tt); International (it); Regional (rc); Hierarchy (h); Market (m); Network (n); Community (c) (blocks – blue circle and cut-points – red square).

itm1, itm2, itm3, itm4 and itn4). The Trinidad and Tobago tourism policy network is particularly susceptible to fragmentation based on the numbers and positions of the cut-points (Fig. 6.2B) as there are more nodes in the Trinidad and Tobago tourism policy network, which can fragment the structure.

Figures 6.3A and 6.3B show the composition of the policy networks for the two fieldwork sites based on Hall's (2011, 2012) typology of Hierarchy, Market, Network and Community. There are a number of observations that can be made from these findings. First, the various types of tourism policy actors are polarized in the Trinidad and Tobago network. From left to right there is a clear separation of the network actors from the market actors with three market actors, ttm1, ttm2 and

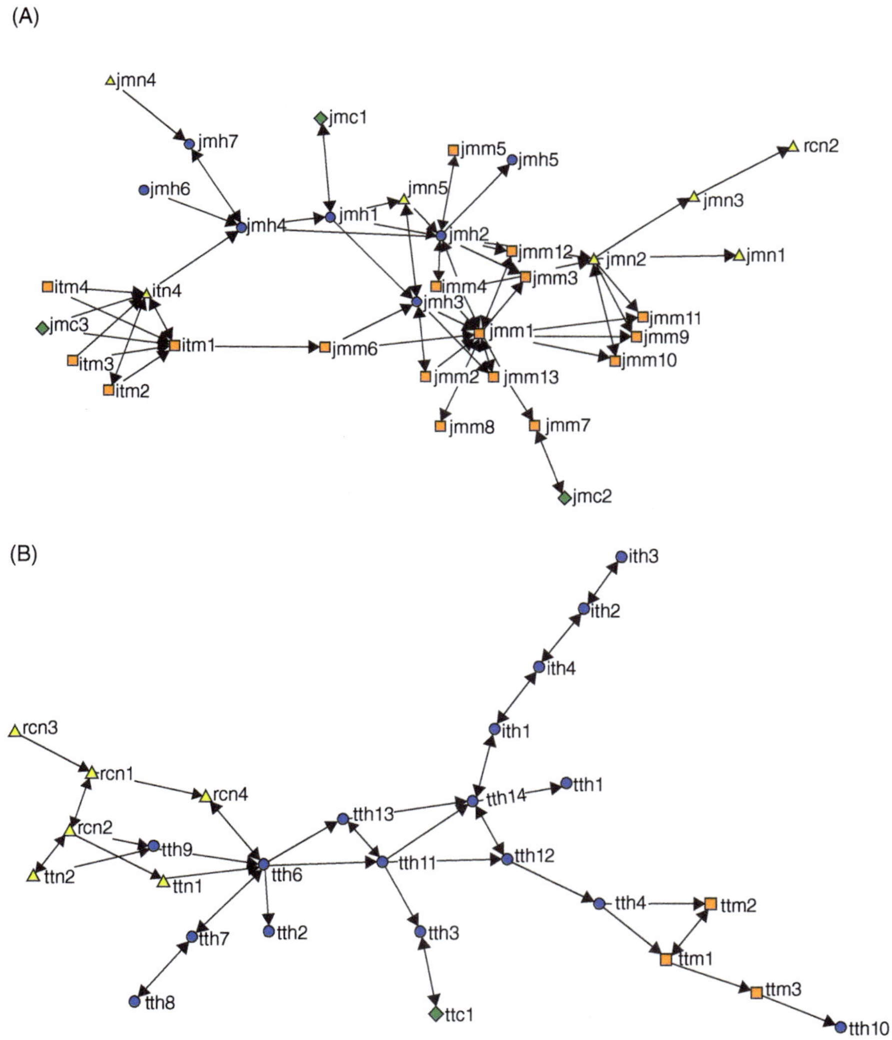

Fig. 6.3. (A) Jamaica tourism policy network (composition). (B) Trinidad and Tobago tourism policy network (composition). Key: Hierarchy (blue, circle); Market (orange, rounded square); Network (yellow, up-triangle); Community (green, diamond).

ttm3, positioned at the far right of the network (Fig. 6.3B). Second, the number of actors in each group is different as while there are 13 hierarchy actors in the Trinidad and Tobago public sector, there are 7 in Jamaica's public sector and for the market actors there are 17 in Jamaica's tourism policy network and 3 in Trinidad and Tobago that can influence tourism policy making. Third, the composition differs in terms of the mix of international, regional and local actors as there are 5 international actors (itm1, itm2, itm3, itm4 and itn4) shown at the bottom left of Fig. 6.3A in the Jamaica tourism policy network and 4 international actors (ith1, ith2, ith3 and ith4) shown at the top right in Trinidad and Tobago tourism policy network (Fig. 6.3B).

Another view of the policy making process involves the institutions and the policy issues being addressed. Institutional actors involved in addressing tourism policy issues for Jamaica and Trinidad and Tobago were identified in McLeod and McNaughton (2016) and the stakeholder maps illustrate the inter-relationships between the key stakeholders, tourism data and the main tourism policy issues in five Caribbean countries. The institutional stakeholder maps for Jamaica and Trinidad and Tobago show different characteristics as different institutions address various policy issues (McLeod, 2015; McLeod and McNaughton, 2016). For instance, in Jamaica the Tourism Product Development Company handles capacity management and manages environmental issues and the Jamaica Tourist Board addresses marketing issues, whereas in Trinidad and Tobago a single entity, the Tourism Development Company, handles and manages all these issues. These institutional distinctions influence the policy formulation process as coordination of the actors' interactions between the institutions would be rather important to ensure the policy making process works.

6.5 Reconstituting the Tourism Policy Making Process

For successful tourism policy making in both island nations there is need to consider the composition of the policy networks and the goals that have been articulated in the policy documents of both countries. Although there are on average 30.5 actors in each network there are differences in terms of the governance actors' inputs into the policy making process as the composition of the actors vary and policy making in the tourism sector falls largely within the influence provided by one particular group of governance actors in the policy networks. Assessing differences in the composition of policy networks are important to bring about the right balance in the policy making process to achieve policy goals and also to understand the distribution of influence among the policy actors in the policy making process.

In terms of the policy networks in the two Caribbean countries, certain agencies and the nature of the links are more important for the formulation of tourism policy. For example there is an evident dominance of market actors in the Jamaica policy network and hierarchy actors in the Trinidad and Tobago policy network. Clearly, this should result in policies being formulated to reflect on the role played by market actors in tourism development in Jamaica as there are 17 market actors in the Jamaica tourism policy network (Fig. 6.3A) as compared to 3 market actors in the Trinidad and Tobago tourism policy network (Fig. 6.3B). Nevertheless, the stated tourism policy in Jamaica seems to be focused more on the benefits that are derived by locals from the tourism sector. In addition, in Trinidad and Tobago the number and position of cut-points in the tourism policy network (Fig. 6.2B) can potentially stall the policy making process as the actors have weak links that lack consensus building. Another concern is that tourism policy making in Trinidad and Tobago is primarily public sector driven. Collaboration and partnering with market actors to a greater extent as suggested by Dredge (2006) should be addressed in Trinidad and Tobago's tourism policy making.

The divergence between stakeholder interactions and stated policy goals should be addressed by reconstituting the policy making process. Without reconstitution, the typologies of both policy networks illustrate potential avenues for tourism policy failure particularly if a wide range of stakeholders is not involved in the policy making process (see Fayos-Solá, 1996). For example, the number of community actors has resulted in the limited advancement of community-based tourism and its role in poverty alleviation and might also have restricted the 'trickle down' effect in tourism. In the case of Jamaica, a Community Tourism Green Paper has been tabled and now a White Paper (Ministry of Tourism and Entertainment, 2015). To ensure that policies are being formulated and implemented, the policy actors that are relevant to the policies should be involved in the policy formulation process and therefore more community actors would be needed in the policy network.

6.6 Conclusion and Lessons Learned

This chapter provides a network perspective of Caribbean tourism policy making drawing on social network analysis, which it is argued can be adapted for political analyses. Using a whole network approach, the formal actors and influences in tourism policy formulation have been clarified. The findings also illustrate that the relationships forged between tourism policy actors are official reporting relationships, inter-board relationships and information sharing relationships, and these are important for the formulation and implementation of tourism policy. The overall network structure of the policy actors and the composition of the network contribute to the formulated tourism policies in the two countries. The goals set out for tourism development to occur should work hand in hand with the policy network characteristics to bring about successful tourism policy formulation and implementation. In the case of both countries there is an opportunity to incorporate policy actors that can assist in achieving the vision for tourism in both countries. An example of this will be including these actors in formal tourism relationships such as Board of Directors.

The contributions of this chapter are that it provides both theoretical and methodological approaches for understanding the tourism policy making process. In this regard, policy making is a process that involves certain inputs of policy actors and an output of articulated policies in policy statements. In both cases, the low influence of community actors belies the governments' commitment to sustainable, responsible and more inclusive tourism development as enumerated in their tourism policy documents. In order to achieve this vision for tourism development there is a need for both countries to seek to develop and implement strategies that can empower communities so that they can potentially have greater influence in tourism policy making and by extension, tourism development. With limited community involvement Jamaica's tourism vision for an 'inclusive, world class, distinctly Jamaican" industry' (Planning Institute of Jamaica, 2009, p. 1) can be challenged.

Tourism policy mechanisms exist within destinations; however, an elaboration of the characteristics of these networks allow for understanding how changes in terms of tourism policy would likely occur and who are the actors involved. This is important as tourism policy can affect tourism development (Jenkins, 1991) and the performance of the tourism sector. An assumption of SNA is that network structures are dynamic and will alter over time as actors, resources and the wider sociopolitical environment within which actors operate change. Policy networks also continue to evolve as, for example, the Ministry of Tobago Development has now been closed (TobagoNews, 2015) and therefore actor tth13 is no longer in existence. In this context, it would be useful for future research to analyse the evolution of tourism policy networks by conducting longitudinal studies. In particular, the involvement of agency actors from other governance types can be tested to determine how the policy making process might change.

Acknowledgements

The authors acknowledge the support of the Faculty of Social Sciences, The University of the West Indies, Mona Campus, Jamaica.

References

Airey, D. and Ruhanen, L. (2014) Tourism Policy-Making in Australia: a national and state perspective. *Tourism Planning and Development* 11(2), 149–162.

Borgatti, S.P. (2002) *Netdraw Network Visualisation*. Analytic Technologies, Harvard, Massachusetts.

Borgatti, S.P. and Halgin, D. (2011) On network theory. *Organization Science* 22(5), 1168–1181.

Borgatti, S.P., Everett, M.G. and Freeman, L.C. (2002) *UCINET 6 for Windows: Software for Social Network Analysis*. Analytic Technologies, Harvard.

Borgatti, S.P., Everett, M.G. and Johnson, J.C. (2013) *Analyzing Social Networks*. SAGE Publications Ltd, London.

Caribbean Tourism Organization (2014a) Country Statistics. CTO, Barbados. Available at: www.onecaribbean.org (accessed 1 September 2014).

Caribbean Tourism Organization (2014b) *Caribbean sustainable tourism policy framework*. Caribbean Tourism Organization, Barbados.

Caribbean Tourism Organization (2015) Caribbean Tourism Review, industry update – end of year 2014. Caribbean Tourism Organization, Barbados.

Central Statistical Office (2014) Visitor arrivals for carnival periods 2000–2012. Central Statistical Office, Trinidad and Tobago.

Dela Santa, E. (2013) The politics of implementing Philippine tourism policy: a policy network and advocacy coalition framework approach. *Asia Pacific Journal of Tourism Research* 18(8), 913–933.

Dredge, D. (2006) Policy networks and the local organisation of tourism. *Tourism Management* 27(2), 269–280.

Dredge, D. and Pforr, C. (2008) Policy networks and tourism governance. In: Scott, N., Baggio, R. and Cooper, C. (eds) *Network Analysis and Tourism: From Theory to Practice*. Channel View Publications, Clevedon, UK, pp. 58–78.

Dye, R.T. (2008) *Understanding Public Policy*, 12th edn. Pearson/Prentice Hall, Englewood Cliffs, New Jersey.

Fayos-Solá, E. (1996) Tourism policy: a midsummer night's dream? *Tourism Management* 17(6), 405–412.
Hall, C.M. (2011) A typology of governance and its implications for tourism policy analysis. *Journal of Sustainable Tourism* 19(4–5), 437–457.
Hall, M. (2012) A typology of governance and its implications for tourism policy analysts. In: Bramwell, B. and Lane, B. (eds) *Tourism Governance: critical perspectives on governance and sustainability*. Routledge, Oxford, UK.
Hall, M. and Jenkins, J. (2004) Tourism and Public Policy. In: Lew, A.A., Hall, C.M. and Williams, A.M. (eds) *A Companion to Tourism*. Publishing Ltd, Blackwell, UK, pp. 523540.
Jamaica Information Service (2015) Ground to be broken this year for 1,600 new hotel rooms. Available at: http://jis.gov.jm/ground-to-be-broken-this-year-for-1600-new-hotel-rooms/ (accessed 11 October 2015).
Jenkins, C.L. (1991) Tourism development strategies. In: Bodlender, J., Jefferson, A., Jenkins, C. and Lickorish, L. (eds) *Developing Tourism Destinations, Policies and Perspectives*. Longman, Harlow, UK, pp. 61–77.
Kenis, P.N. and Schneider, V. (1991) Policy networks ad policy analysis: scrutinizing a new analytical toolbox. In: Marin, B. and Mayntz, R. (eds) *Policy networks: empirical evidence and theoretical considerations*. Westview Press, Boulder, Colorado, pp. 25–29.
Knoke, D. and Yang, S. (2008) *Social network analysis*, 3rd edition. SAGE, London.
McLeod, M. (2015) *Caribbean Tourism Sector Open Data Scoping Study*. Caribbean Open Institute, Kingston, Jamaica.
McLeod, M. and McNaughton, M. (2016) Mapping an emergent Open Data eco-system. *Journal of Community Informatics* 12(2), 26–46.
Ministry of Tourism (2010) National Tourism Policy of Trinidad and Tobago, Trinidad and Tobago.
Ministry of Tourism and Entertainment (2015) National Community Tourism Policy and Strategy. Jamaica: Ministry of Tourism and Entertainment.
Mizruchi, M.S. (1994) Social network analysis: recent achievements and current controversies. *Acta Sociologica* 37, 329–343.
Otte, E. and Rousseau, R. (2002) Social network analysis: a powerful strategy also for the information sciences. *Journal of Information Science* 28(6), 441–453.
Pforr, C. (2006) Tourism policy in the making. *Annals of Tourism Research* 33(1), 87–108.
Pforr, C., Pechlaner, H., Volgger, M. and Thompson, G. (2014) Overcoming the limits to change and adapting to future challenges: governing the transformation of destination networks in Western Australia. *Journal of Travel Research* 53(6), 760–777.
Planning Institute of Jamaica (2009) Vision 2030 Jamaica Final Draft Tourism Sector Plan, Jamaica.
Rhodes, R.A.W. (2006) Policy Network Analysis. In: Moran, M., Rein, M. and Goodin, R.E. (eds) *The Oxford Handbook of Public Policy*. Oxford University Press, Oxford, UK, pp. 425–447.
Ritchie, B.J.R. and Crouch, G.I. (2003) *The Competitive Destination: A Sustainable Tourism Perspective*. CAB International, Wallingford, UK.
Saxena, G. (2005) Relationships, networks and the learning regions: Case evidence from the peak district national park. *Tourism Management* 26(2), 277–289.
Scott, J. (2013) *Social network analysis*, 3rd edn. SAGE, London.
Stevenson, N., Airey, D. and Miller, G. (2008) Tourism policy making: the policymakers' perspectives. *Annals of Tourism Research* 35(3), 732–750.
Strobl, A. and Peters, M. (2013) Entrepreneurial reputation in destination networks. *Annals of Tourism Research* 40, 59–82.
TobagoNews (2015) It is official; Ministry of Tobago Development folds. Available at: http://thetobagonews.com/eng-en/index.php/home/headlines/item/808-it-is-official-ministry-of-tobago-development-folds.
Tyler, D. and Dinan, C. (2001) The role of interested groups in England's emerging tourism policy network. *Current Issues in Tourism* 4(2–4), 210–252.

Wasserman, S. and Galaskiewicz, J. (1994) *Advances in social network analysis: research in the social and behavioural sciences.* SAGE, London.

Wong, E.P.Y., Mistilis, N. and Dwyer, L. (2010) Understanding ASEAN tourism collaboration – the preconditions and policy framework formulation. *Tourism Management* 12(3), 291–302.

Appendix I

Trinidad and Tobago Tourism Policy Network Actors

Hierarchies (h)	Markets (m)	Networks (n)	Communities (c)
Chaguaramas Development Authority Central Statistical Office Environmental Management Authority Evolving Technologies and Enterprise Development Company Intergovernmental Panel on Climate Change Tourism Development Company Limited Tobago House of Assembly Division of Tourism and Transportation Tobago Hospitality and Tourism Institute Trinidad and Tobago Hospitality and Tourism Institute Urban Development Corporation of Trinidad and Tobago Limited Ministry of Tourism Ministry of Trade and Industry Ministry of Tobago Development Ministry of Planning and Sustainable Development United Nations World Tourism Organization World Health Organization United Nations	Hilton Trinidad and Conference Centre Magdalena Grand Hyatt Hotel	Caribbean Alliance for Sustainable Tourism Caribbean Hotel Association Caribbean Tourism Development Company Caribbean Tourism Organisation Tobago Hotel and Tourism Association World Travel and Tourism Council Trinidad Hotels Restaurants & Tourism Association	Turtle Village Trust

Appendix II

Jamaica Tourism Policy Network Actors

Hierarchies (h)	Markets (m)	Networks (n)	Communities (c)
Ministry of Tourism and Entertainment / Tourism Enhancement Fund/ Jamaica Vacations Limited Jamaica Tourist Board Tourism Product Development Company Limited Ministry of Transport and Works/National Works Agency Jamaica National Heritage Trust Norman Manley International Airport Port Authority of Jamaica	Sandals Resorts International Superclubs Chukka Caribbean Adventures Jamaica Tours Limited Rondell Village Sunflower Resort and Villas (Fisherman's Point Hotel, Skycastles) Jakes Hotel, Villas & Spa Jamaica Co-operative Automobile and LimousineTours Ltd. Virgin Atlantic Can jet Sunwing Delta Airlines Courtleigh Hotel & Suites Holiday Inn, Jamaica Pegasus, Knutsford Court Hotel) (JTB & JHTA) Dolphin Cove (JHTA) Round Hill Hotel & Villas (JHTA) Tensing Pen (JTB) Island Routes (TPDCo)	Jamaica Protected Areas Trust/Forest Conservation Fund Jamaica Hotel and Tourist Association Jamaica Union of Travellers Association Association of Jamaica Attractions Limited Jamaica Association of Villas and Apartments International Institute for Peace through Tourism International Air Transport Association Interntional Civil Aviation Organization Caribbean Hotel Association	Community Tourism Partners Committee BREDS The Treasure Beach Foundation Countrystyle and Unique Jamaica

7 Planning for Growth in Islands: The Case of Cuba

RACHEL DODDS,[1]* FREDERIC DIMANCHE[1] AND MICHAEL SADOWSKI[2]

[1]*Ryerson University, Toronto, Canada;* [2]*Intrepid Group North America, Vancouver, Canada*

7.1 Introduction

The characteristic complexities of island destinations give rise to many resource management and governance issues (Douglas, 2006). Islands are unique in that they have firmly established boundaries because of the limitations of water surrounding them, yet they often also maintain a feeling of relaxation and tranquillity that landlocked nations lack, possibly because they require different modes of transportation to get to. Many islands have autonomy due to their own management and political structure, yet they are often extremely vulnerable to pressure from development and other human activities. This chapter examines challenges and successes of island development and then examines the island of Cuba outlining future issues that this island may face as it is in the midst of development.

Whether in warm or cold water, many different forms of tourism exist in islands. In warm-water islands, resort or beach tourism is often the most prevalent (e.g. Caribbean and Mediterranean islands), yet some can also be centres for commerce and banking (e.g. Cayman Islands). Mass tourism and low-end all-inclusive resorts can exist in some (e.g. the Dominican Republic or Mallorca) or upscale hideaways that cost thousands per night can dominate (e.g. Mustique Island in the West Indies has a price tag of $24,000/night). Islands can host crowded city experiences (e.g. Hong Kong or Singapore) or rural uncrowded experiences (e.g. Bora Bora or Dominica). Some islands boast private experiences (e.g. you must rent the entire island of Musha Cay in the Bahamas), while others offer a community experience (e.g. Rio Esteban in Northern Honduras).

There are a number of approaches for island tourism development. Community-based tourism (CBT) centres on ownership, management and control of tourism projects by the local community (Simpson, 2008) and some islands such as Cape Verde (Cañizares, Castillo Canalejo and Núñez Tabales, 2015), focus on this approach to develop tourism while others focus more on the participatory, multi-level

*E-mail: R2dodds@ryerson.ca

governance for CBT such as the Galapagos (Ruiz-Ballesteros and Brondizio, 2013). On the other end of the spectrum is resort tourism where developers do not involve local inhabitants at all, except for some, usually menial, employment opportunities. Resort islands are equipped 'with comprehensive facilities for accommodation, food, recreation and leisure. They are also strictly reserved for foreign tourists and guarantee complete privacy' (Domroes, 2008, p. 122). Some islands such as the Maldives (Domroes, 2008), or Brijuni in the Mediterranean (Urošević, 2014), focus on this type of resort tourism as a form of development. Other islands such as Mallorca (Louisa, 2013) or Dominica (Patterson, Gulden, Cousins and Kraev, 2004), that do not possess beaches or do not wish to promote beaches as the main attraction may focus on ecotourism or nature-based tourism.

When the appeal of some island destinations is diminished due to tourist crowds (e.g. Balearics, Greek Islands, Cyprus and the Canaries in Europe, as well as some islands in the West Indies and South Pacific), some islands are switching to more community-based approaches such as Patong Beach in Phuket (Polnyotee and Thadaniti, 2015) to offset negative impacts experienced. Other islands offer multiple forms of tourism such as St. Lucia who use both resort tourism and community-based approaches (Dodds, Ali and Galaski, 2016) or Jamaica (Jayawardena, 2002).

7.2 Tourism Planning and Development Strategies: Challenges

Many islands depend heavily on the natural resources within their destination and it is these resources that their industries have traditionally been based upon. The need to plan for tourism on islands is especially crucial given their limited size and resource base (Pearce, 2008). As the sustainability of many destinations' economies is now dependent on tourism, the need for planning and management of a destination is imperative. Tourism can be beneficial in many ways. It provides jobs, facilitates infrastructure and can showcase and preserve a destination's cultural and heritage attractions. Unfortunately, with the growth of tourism, not all development strategies have been sustainable. Sustainability can be described as development that considers how destinations can be contained, preserved and managed addressing the triple bottom line: economic, social and environmental.

There are typically seven challenges that islands face with regard to planning:

1. The first challenge is infrastructure. Often local infrastructure is not adequate for the additional numbers of temporary residents an island may attract. According to Graci and Dodds (2010, p. 40) 'there are often inadequate sewage facilities and this has damaged mangroves, polluted lagoons and degraded fragile reef systems.' The continuous focus on increasing tourism numbers rather than focusing on yield has led to construction of large-scale infrastructure projects such as airports, cruise ship terminals, golf courses and holiday resorts that often have altered delicate coastlines and defaced mountain sides. Some strategies to attract high-yield tourism have also, at times, turned out badly. Marina development including anchoring, sand mining and sewage discharge has damaged the environment. Quick and unplanned development has resulted in water shortages, electricity blackouts and a lack of the island's ability to dispose of waste (Dodds, 2012; 2013).

2. Another challenge is that of a copycat strategy. Concern over intense regional and international competition leads destinations to copy another product that is based on the development of new product offerings and the exploitation of local resources. For example, in St. Kitts, as in many other islands, a port was built to accommodate large cruise ships to compete with other destinations. This port was built without regard to economic benefits that were not proportional to the level of visitations that resulted (Dodds and McElroy, 2008).

3. Next is a lack of strategic planning. Many islands lack long-term sustainable development strategies and therefore have focused only on short-term benefits (e.g. St. Kitts, Malta). Tofino on Canada's Vancouver Island suffered from not having a strategic plan to respond to issues affecting tourism (such as the 2006 and 2009 water shortages). Although tourism is the main contributor to the economy in Tofino, no formal plan or policy existed and local stakeholders could not identify more than a few sustainable tourism practices (Dodds, 2012). In other islands, government may not have long-term plans as their governance structures are for 4–5 years instead of 10–20, which would allow for planned growth and long-term sustainability goals. Short-term returns often take priority over long-term ones and mass tourism is often developed with the aim of generating short-term gains. Not all impacts are environmental. Some of the most successful tourist destinations have also experienced social disruption (e.g. Malta, Balearics) and cultural losses (e.g. islands of Thailand, Indonesia and Hawaii). Many of these non-sustainable outcomes have been due to the absence of planning considerations for environmental protection and/or heritage of cultural preservation (Dodds and Butler, 2010; Graci and Dodds, 2010). While stakeholders often lack a common consensus about what 'tourism should be', there is often failure of policymakers to manage growth within social and ecological limits.

4. Another challenge is that of environmental awareness on the part of government and lack of coordination and integration of initiatives into wider policy efforts. Lack of awareness of tourism's needs is evident in the failure of many island destinations to hold in check land encroachment by international developers. The following issue in St. Kitts can illustrate an example of lack of integration: 'Although entrepreneurship is encouraged by the Tourist Board, there are no economic incentives or training programs to assist new business development or community and rural development' (Dodds and McElroy, 2008, p. 8).

5. A fifth challenge is that of economic priority over social and environmental concerns. Often, much needed tax dollars are allocated towards the development of transportation linkages such as airports and boat docks. As the only way to get to an island is via plane or boat (with the exception of bridges, occasionally), it is necessary to promote and invest in such accessibility methods. However, many times this is at the expense of much needed infrastructure such as hospitals and schools (Graci and Dodds, 2010). Many islands fall into the misunderstanding that the development of transportation linkages will bring more tourists thus leading to further revenue to be gained for other infrastructure for the community. The reality, however, is that many of these linkages do not benefit all islanders and are often used on a seasonal basis (depending on tourism seasonality), not to mention that tourism is not always growing, so it is possible for islands to place themselves into debt to foreign development banks or governments in order to build these superstructures.

6. One of the biggest challenges that islands face is that natural and physical resources often limit their economic development. Islands, due to their size, may not have sizeable amounts of physical space or natural resources. Often islands must import food and other supplies and water is often scarce. Many small islands are unlikely to be able to rely on underground aquifers or catchments from large rivers for fresh water and this problem is often aggravated by the ratio of island circumference to number of visitors.
7. Finally, many islands face the issue of foreign ownership and resulting economic leakage. Foreign or multinational corporations may own many resorts or accommodation facilities on islands. Cole (2006) identified that businesses that are owned by foreigners do not contribute to the local economy as much as they would have if local residents owned them. Carter (1991) also noted that foreign-owned airlines that operate in developing countries have a great influence on tourist arrivals, in turn influencing the potential for economic growth in certain areas. This results in leakage (where the majority of the revenue leaves the island for imports, savings or payment of foreign workers). In the Caribbean, for example, Belisle (1983) documented leakage due to food imports. Leakage can be as high as 80% in these islands (United Nations Environment Program, n.d.) where tourism development is influenced by foreign investors and by the necessity to import goods and services to satisfy tourists.

7.3 Tourism Planning and Development Strategies: Successes

Seventy years of post-war tourism has taught policy makers some valuable lessons, and many islands are now moving toward better planning and development following a number of successes:

1. First is the implementation of policies. Some successes include highly developed islands like the Spanish Balearics and Malta that are implementing sustainable development policies (Dodds, 2007a; 2007b), while newer resorts like those in the Maldives are initiating environmental assessments and planning for climate change (Ghina, 2003).
2. Another successful approach is the commitment to planning. Bermuda and the Seychelles, both with a tradition of environmental conservation, share a distinct, self-conscious, upscale tourism identity and are committed to long-term planning.
3. Diversification away from a strong dependence on tourism is another success strategy. Islands such as the British Virgin Islands, Cayman Islands, Mauritius and Northern Marianas have avoided excessive tourism dependence by means of diversifying into other profitable economic activities like off-shore financing, sugar, textile manufacturing and power generation (McElroy, 2003). Samoa is also a good example of an island that has undertaken strategic long-term planning (Pearce, 2008). Aruba won the *National Geographic World Legacy* Award Winner for 'Destination Leadership due to their aggressive efforts in renewable energy such as wind and solar' (PR Newswire, 2015).
4. Local ownership is another success. Some of the outer islands of Scotland, Finland, Sweden and even Canada are dominated by locally owned small operators

and often have little external investment keeping all monies locally in the community (Graci and Dodds, 2010).

5. In some cases island circumstances have influenced development. Restricted transport infrastructure has allowed eco-friendly or other types of low-density tourism to flourish. Diving in Bonaire, ecotourism in Dominica or volcanic exploration in Montserrat have allowed for development controlled and owned by locals rather than by global package mass tour operators (Dodds and McElroy, 2008). Accessibility limited by a small airport has prevented the growth of mass tourism in St. Barthélemy, in the leeward Caribbean islands, and has helped to develop an up-market destination for French cuisine (Baldacchino, 2006).

6. Although not a strategy for success, in many isolated cold-water islands like the Shetlands, the Falklands, Antarctica and Greenland, tourism is limited by a short holiday season, expensive access and extreme adventure and exploration that are not widely demanded activities by tourists. This results in limited tourism 'footprint' on cold-water islands (Baldacchino, 2006) and allows for other activities to flourish rather than the sole focus on tourism.

7.4 The Case of Cuba

Cuba as a transition economy

Cuba is a case of an island with an economy in transition. In other words, Cuba is evolving from a centrally planned (communist) economy to a market or mixed market economy, and tourism is contributing greatly to this transition by providing development opportunities for small businesses and entrepreneurship. The tourism sector can be considered to be leading economic reforms and fostering small business development and self-employment. Tourism exposes the locals to foreign travel companies and international travellers who contribute to significantly increase the country's Gross Domestic Product (GDP). Travel and tourism's total contribution to Cuban GDP is forecasted to rise by 4.9% annually to reach 12.3% of GDP in 2026 (World Travel and Tourism Council, 2016). The tourism literature that discusses the issues countries with transition economies face when developing tourism, however, is still quite limited (e.g. Simon, 1995; Williams and Baláz, 2002; Bramwell and Meyer, 2007). Cuba will continue to be an interesting case to document in their efforts to experience what other countries such as China, former Communist European countries or Vietnam have previously experienced.

Since former President Barack Obama restored diplomatic relations with Cuba following a 2015 announcement, there have been endless articles heralding the reduction of travel restrictions for US citizens. Americans are allowed to visit Cuba under 12 broad categories of travel activities once they have obtained a travel visa. Due to this new relationship, Cuba has positioned itself as a cultural and heritage destination for America as well as other international destinations.[1]

As American travel restrictions have been reduced, Cuba's plans are to grow exponentially. For example, there are plans to grow from 60,500 rooms in 2013 to 80,600 rooms by 2020 (Personal communications, 2015). The majority of new growth will be on northeast side of the island, which has 56 virgin beaches such as Caya Sabinal, Caya Cruz, Caya Romano and all new developments will be resort development. The objective for Cuba is to increase tourism arrivals to five million by 2020.

History

The tourism sector faces several unique challenges owing to the socialist economy maintained by the Cuban government and the embargo maintained against the country by the US. Nevertheless, Cuba has increasingly become open to international tourism as a means of procuring foreign capital and stimulating economic growth. The Cuban tourism industry must operate between the bounds of the norms established by generally capitalist neighbouring Caribbean islands that act as competition and government pressures to maintain the socialist political economy of the island nation.

Cuba is a large Caribbean island located 150 kilometres south of Florida, USA, and surrounded by neighbouring Caribbean and Central American countries such as Mexico, the Bahamas, Jamaica, the Cayman Islands and Haiti. It was a Spanish colony from the beginning of European settlement in the New World up until the Spanish–American War of 1898, at which point it became a client state to the United States (CIA World Factbook, 2017). Cuba remained politically and economically dependent on the USA until 1961 and, by the 1950s, was 'one of the four or five most developed nations in Latin America, and the most developed tropical nation in the entire world' (Foran, 2009, p. 18). The principal industries that contributed to this favourable economy were agricultural with sugar being the most important export industry. Sugar amounted to 80% of Cuba's exports at this time, and Cuban sugar comprised more than half of the world market (Foran, 2009). While Cuba was economically prosperous at this time, the country was economically and politically stratified, with the dictatorial government of Fulgencio Batista enforcing the interests of the capitalist elite within the country.

The Cuban Revolution overthrew this neo-colonial social economy and established a socialist government under the rule of Fidel Castro and his Communist Party. Castro exploited nationalist discontent at American dominance as well as populist sentiment directed against Batista's elitist government (Foran, 2009). The late 1950s saw an economic downturn in Cuba that came suddenly, and this further undermined the legitimacy of the Batista government. In this regard, Cuba experienced a transformation of its key industries that paralleled many of its neighbours, albeit in a more revolutionary fashion. Most Caribbean nations were structured similarly to Cuba prior to its revolution, in that they were politically and socially linked to a colonial power in Europe or North America and had their economies organized around agricultural exports. International tourism throughout the Caribbean at this time comprised only a small proportion of Gross Domestic Product (GDP) and was primarily oriented towards only the wealthiest travellers from North America and Europe. By the 1960s, agricultural exports were proving a less reliable means of economic growth throughout the Caribbean and governments in the region attempted to modernize and diversify their economies. The development of mass international jet travel created the potential for tourism to contribute to local economies to a greater extent and many countries took advantage of their natural beauty to promote this activity.

Cuba did not develop a mass tourism industry in the same way that its neighbours did, at least not during the first 30 years following the revolution. Instead, Cuba developed a socialist economy. By the late 1980s, Cuba had a negligible unemployment rate and spent more on healthcare and education than any nearby country with a comparable economy (Foran, 2009). Even today, government consumption accounts for 34.4% of GDP in an economy where 73.9% of GDP originates in the service sector (CIA World Factbook, 2017). Prior to the 1990s though, Cuba's economy was effectively buoyed up by its relationship with the Soviet Union. The collapse of the Soviet Union exposed the extent to which Cuba relied on Soviet trade and aid: Cuba's gross national product plummeted by 40% between 1989 and 1992 and Cuba was forced to reorganize its economy in the absence of a foreign patron (Foran, 2009). Cuba was able to avoid economic and political collapse, however, through a remarkable period of economic diversification in which tourism played an increasingly significant role.

Tourism was vital to the economic restructuring of Cuba but has come to represent just one pole in a more diversified economy since the crisis of the 1990s. After the end of Soviet aid, tourism grew dramatically in Cuba from the almost non-existent place it occupied before. It overtook sugar exports as the largest foreign currency earner in 1997 when it accounted for 43% of GDP (Wilkinson, 2008). But the economic recovery experienced in the 1990s relied on diversification as much as tourism, and Cuba now has significant industries in biotechnology and oil exploration as well as in mining, construction, agricultural manufacturing and agricultural exports (CIA World Factbook, 2017). Tourism is therefore relatively less important to the contemporary Cuban economy than it was during the heydays of the 1990s while remaining an important sector in a diversified economy.

Cuba's attractiveness

Cuba is a country rich in natural resources that comprise 400 beaches and 500 km of white-sand coastline. Approximately 25% of the country is designated as a protected area and there are seven biosphere reserves and seven national parks. Cuba also has a rich history and heritage. The island boasts 10 World Heritage Sites, 332 museums and 257 national monuments. Eight of Cuba's cities are over 800 years old and there are 60,000 classic cars. Cuba is a country that is politically stable and there are little drugs and no guns. It is known as a safe destination with excellent health care and one of the world's highest literacy rates.

Cuba is an island that is dependent on tourism and also fuelled by tourism. Since the American economic embargo starting in the 1950s and the fall of the Soviet Union, Cuba has depended on tourism for much of its income and its largest trading partners are Canada, Spain, Venezuela and China. The past 20 years of tourism have accounted for 46% of the incoming dollars into Cuba and visitation increased 9.8% year over year between 1990 and 2014. The number of international tourists reached 3.3 million in 2015. Cuba is the third most popular international destination for

Canadians who represent the current majority of tourists to the island. There are currently 60,000 rooms in Cuba, up from 18,000 in the 1990s, and current occupancy levels are about 60%. It is also one of the few countries that do not trade with the USA (although this is due to change with the 2015 Cuba–USA agreement). Due to the Cuba Clause, any ship that touches Cuban land cannot visit the USA within 6 months. This clause has limited procurement and imports drastically, therefore increasing the dependency on tourism.

Cuba is different from many other destinations as there is no foreign ownership in Cuba: all ownership must follow the 49–51% rule. Due to this stipulation, four major Cuban hospitality companies, Cubacan, Gran Caribe, Azul and Gaviotas, all share a major portion of their business with foreign enterprises (Gaviotas, for example, holds 32% of all tourism businesses in Cuba even though many of them may be owned by international chains such as Melia, Iberostar, etc.).

The island is poised for a strong increase in demand. The geographic proximity to the US market in general, and to the numerous Cuban–Americans who have been waiting for years for opportunities to visit their former country and heritage constitutes the main growth factor. The first American cruise ship to visit Cuba from the USA in nearly 40 years crossed the Florida Straits from Miami Havana on May 2nd, 2016, as a symbol to the end of a cold war era (Havana Times, 2016). More recently, the first American passenger plane to complete a commercial flight to Cuba in 50 years landed in La Havana on August 31st 2016. The challenge for Cuba will be to respond to demand, control development and develop the necessary supply at quality levels that American and other travellers will increasingly demand.

Policies

Current policies are governing the development of tourism in Cuba. The following identifies some of the key policies with respect to business ownership and environmental planning.

Ownership:

- No foreigner can own property in Cuba; however, many hotel chains have 51–49% joint ventures.
- Independent home stays called *casas* used to be taxed $140 per month every month and charged two months in advance, which could be prohibitive for small owners, especially in the low season. Regulations have now relaxed and the tax is $40–50 per month without a limit to the number of rooms per *casa*.

Environmental planning:

- All new hotels must have solar panels and all new hotels in tourist areas must use solar energy.
- To help with water shortages that are becoming ever more present, the government is planning to have a channelling system for water from the mountains and build dams.
- Construction is restricted to 50 meters away from the shoreline and there are regulations against building on sand dunes.

- Developers cannot build higher than two stories above palm-tree height. The *local office of the historian* enforces building height restrictions.
- All fertile land near new resort areas (except Varadero) will grow local produce such as capsicums, lettuce, tomatoes, etc. to supply resorts.
- Before any new hotel is built, an environmental assessment must be done by a national Cuban organization called SIGMA.
- San Cristobal is an autonomous agency in Cuba that has to date restored 33% of all buildings in old Havana. There are plans to continue with heritage restoration in other towns in Cuba.

Sustainability issues

The sustainability of Cuba as a tourism destination remains questionable. All planned growth is for all-inclusive resort development even though Cuba's unique selling proposition and current marketing strategy is that of *Auténtica Cuba,* which is focused on heritage: history, culture and people. With Cuba still being under the effect of an economic embargo from the USA, another major issue will be the procurement of goods necessary to cater to increasing tourist needs. Large resorts as well as small private enterprises such as the *casas* (independent homes for rent) are likely to continue to be impeded from the embargo. This may affect the quality of service that international guests expect.

In addition, there is uncertainty with respect to the political future of Cuba. The current leader, Raúl Castro, has announced he will not be serving another term after 2019. This will be a turning point for Cuba as there must be a consensus within government to decide how many terms politicians will then serve. Although the transition towards a mixed market economy is unlikely to be stopped, a stable political situation is essential for investor confidence and economic growth as well as to plan for long-term sustainable development policies. Along with political change, major infrastructure issues remain. Throughout the country, there are currently no plans for water conservation even though Cuba is facing significant water issues. The only plans are to impact nature (building dams and waterways) rather than adapting water consumption behaviour. A successful and sustainable island tourism destination should be particularly attentive to managing its water resources. Clean (drinking) water is necessary for safe consumption by locals and tourists, and clean (sea) water is essential to Cuba as an attractive beach destination. An effort to assess risk for water resources should be a priority for environmental sustainability (Roe, Hrymak and Dimanche, 2014).

In the capital Havana, which 70% of tourists visit, they are planning to re-open a harbour that will be partly a ferry access to and from Miami and partly a cruise terminal. As Havana currently has few motor vehicles and still has horses and buggies, an increase in motorized vehicle traffic could be catastrophic if not properly planned and controlled together with road infrastructure development. Trinidad, located southeast of La Havana, is another area facing significant infrastructure issues. It is one of the fastest growing destinations in terms of visitation, but it has poor sewage, few accommodation properties apart from home stays, no parking and water scarcity issues.

Winson (2006) conducted one of the few studies that investigated sustainable ecotourism practices in Cuba and the impact of political constraints to tourism sustainability. Clearly, the jury is still out. Cuba can be considered as a destination torn between inward looking stagnation inherited from the political system and the desire to grow an international economy exemplified by tourism development. As Rodríguez, Parra-López and Yanes-Estévez (2008, p. 63) suggest, 'the adoption of sustainability strategies must include measures for the conservation and protection of the environment, as well as land use planning in general. If these strategies are to have a positive impact on the environment, they must incorporate a regulatory framework in relation to the environment.' Cuba's success as a sustainable tourism destination will be dependent upon coordinated tourism development, land use planning and an environmental regulatory framework.

7.5 Conclusion

Although every destination must shape its own path, a generation of island tourism experience suggests that success can come from:

- proactive and strategic planning with a long-term time horizon;
- public environmental education and community participation in tourism decision-making;
- economic development diversification beyond tourism dependence; and
- broad policy acceptance on limitations to growth.

Achieving these elements, in whole or in part, will ensure islands are moving towards profitable tourism that is socially acceptable to the host population and environmentally sustainable for future generations. For island tourism to be successful, careful consideration must be given at the outset to determining just how much tourism development is desired and feasible. Examining Cuba, there are a number of enforced environmental policies that are often lacking in other island destinations. However, the current limitations to growth and the type of tourism planned for Cuba's immediate future question the very nature of sustainable development.

This chapter argues that for island tourism, as any other destination, planning must be considered if a competitive, sustainable tourism industry is to ensue. From this argument stems a number of lessons that can be learned and used as a road map for moving forward. First, while it is too early to determine whether Cuba will continue on its path for extreme growth, both industry and government should be aware of the impacts of quick growth as demonstrated in many other islands globally. Second, Cuba is a unique island in that it has environmental and social policies that many other islands are lacking. The island demonstrates some forward thinking for renewable energy used in tourism as well as a focus on cultural restoration that is focused on the benefit for the local population as well as tourists. Putting the well-being of the community first is a positive approach to sustainable tourism planning and Cuba should be commended for its efforts in this regard. Finally, although many other islands are not in the same political space that Cuba is, they are still heavily influenced by outsiders. In the case of Cuba, the influence of the USA should not be considered lightly.

Cuba is on the verge of change much faster than most other islands. Restrictions on Cuba are being lifted and US investors and travellers are bound to have a significant impact, for better or for worse, on the island. Cuba has specific development challenges that are constrained by its history, its geography and its current political situation. It is up to Cuba to seize these opportunities with a careful approach to tourism planning and economic development. Without proper planning and development, Cuba runs the risk of destroying the very heritage it has sought to protect for the past decades. As Sanchez and Adams (2008) pointed out, tourism in Cuba presents a Janus-faced character, with contradictions resulting from the desire to develop economically and the efforts to build an independent nation. It will be important for tourism and socio-economic development researchers to monitor the evolution of Cuba in the next years. Sharpley and Knight's (2009, p. 253) pessimistic prediction that 'although the potential opening of the US tourist market to Cuba would undoubtedly bring an upsurge of demand, the full potential of access to Cuba for US citizens ... is unlikely to be realised' is to be tested. Our hope is that Cuba becomes a champion for heritage and natural preservation, while benefiting from the growth of international tourism.

End note

[1] See www.treasury.gov/resource-center/sanctions/Programs/Documents/cuba_faqs_new.pdf

References

Baldacchino, G. (2006) Warm versus cold water island tourism: A review of policy implications. *Island Studies Journal* 1(2), 183–200.

Belisle, F.J. (1983) Tourism and food production in the Caribbean. *Annals of Tourism Research* 10(4), 497–513.

Bramwell, B. and Meyer, D. (2007) Power and tourism policy relations in transition. *Annals of Tourism Research* 34, 766–788.

Cañizares, S.M., Castillo Canalejo, A.M., and Núñez Tabales, J.M. (2015) Stakeholders' perceptions of tourism development in Cape Verde, Africa. *Current Issues in Tourism* 9(10): 966–980.

Carter, T.R. (1991) *The Hatch Index of Climatic Favourability*. Finnish Meteorological Institute, Helsinki, Finland, p. 18.

CIA World Factbook (2017) Cuba. Available at: https://www.cia.gov/library/publications/the-world-factbook/geos/cu.html (accessed 9 October 2017).

Cole, S. (2006) Information and empowerment: The keys to achieving sustainable tourism. *Journal of Sustainable Tourism* 14(6), 629–644.

Dodds, R. (2007a) Sustainable tourism and policy implementation: Lessons from the case of Calviá, Spain. *Current Issues in Tourism* 10(1), 296–322. DOI: 10.2167/cit278.0.

Dodds, R. (2007b) Malta's tourism policy: Standing still or advancing towards sustainability? *Island Studies Journal* 2(1), 47–66.

Dodds, R. (2012) Sustainable tourism: A hope or a necessity? The case of Tofino, British Columbia, Canada. *Journal of Sustainable Development* 5(5), 54–64.

Dodds, R. (2013) Will tourists pay for a healthy environment? Assessing visitors' perceptions and willingness to pay for conservation and preservation in the island of Koh Phi Phi, Thailand. *International Journal of Tourism Anthropology* 3(1), 28–42. DOI: 10.1504/IJTA.2013.054407.

Dodds, R. and Butler, R.W. (2010) Barriers to implementing sustainable tourism policy in mass tourism destinations. *Tourismos* 5(1), 35–54.

Dodds, R. and McElroy, J. (2008) St. Kitts at a crossroads. *ARA Journal of Travel Research* 1(2), 1–10.

Dodds, R., Ali, A. and Galaski, K. (2016) Mobilizing knowledge: determining key elements for success and pitfalls in developing community-based tourism. *Current Issues in Tourism*, 1–22.

Domroes, M (2008) Conceptualising state-controlled *resort islands* for an environment-friendly development of tourism: The Maldivian experience. *Singapore Journal of Tropical Geography* 22(2), 122–137.

Douglas, C.H. (2006) Small island states and territories: Sustainable development issues and strategies – challenges for islands in a changing world. *Sustainable Development* 14, 75–80.

Foran, J. (2009) Theorizing the Cuban revolution. *Latin American Perspective*, 36(2), 16–30.

Ghina, F. (2003) Sustainable development in small island developing states: The case of the Maldives. *Environment, Development and Sustainability* 5(1), 139–165.

Graci, S.R. and Dodds, R. (2010) *Sustainable Tourism in Islands*. Earthscan Press, London.

Havana Times (2016) *Cuba Gets Cruise Tourism from the USA*. Available at: www.havanatimes.org/?p=118468 (accessed 9 October 2017).

Jayawardena, C. (2002) Mastering Caribbean tourism. *International Journal of Contemporary Hospitality Management* 14(2), 88–93. DOI: 10.1108/09596110210419273.

Louisa, W. (2013) *Nature-based Tourism in Mallorca's Natural Areas: The Benefits of Tourism for Natural areas*. Springer, Wiesbade, Germany.

McElroy, J.L. (2002) The impact of tourism in small islands: A global comparison. In: di Castri, F. and Balaji, V. (eds) *Tourism, Biodiversity and Information*. Backhuys, Leiden, the Netherlands, pp. 151–167.

McElroy, J.L. (2003) Tourism development in small islands across the world. Geografiska Annaler. *Human Geography* 85B(4), 231–242. DOI: 10.1111/j.0435-3684.2003.00145.x.

McElroy, J. and Dodds, R. (2007) What does sustainable tourism mean for islands [Special issue]. *ID21 Island Insights* p. 3.

Patterson, T., Gulden, T., Cousins, K. and Kraev, E. (2004) Integrating environmental, social and economic systems: A dynamic model of tourism in Dominica. *Ecological Modelling* 175(2), 121–136.

Pearce, D.G. (2008) Tourism planning in small tropical islands: Methodological considerations and development issues in Samoa. *Études caribéennes, 9–10*. Available at: https://etudescaribeennes.revues.org/1393?lang=fr (accessed 29 April 2016).

Polnyotee, M. and Thadaniti, S. (2015) Community-based tourism: A strategy for sustainable tourism development of Patong Beach, Phuket Island, Thailand. *Asian Social Science* 11(27), 9098.

PR Newswire (2015) *Aruba Wins National Geographic World Legacy Award for Leading Sustainability Efforts*. Available at: http://www.prnewswire.com/news-releases/aruba-wins-national-geographic-world-legacy-award-for-leading-sustainability-efforts-300046235.html (accessed 1 August 2016).

Rodríguez, J.R.O., Parra-López, E. and Yanes-Estévez, V. (2008) The sustainability of island destinations: Tourism area life cycle and teleological perspectives. The case of Tenerife. *Tourism Management* 29(1), 53–65.

Roe, P., Hrymak, V. and Dimanche, F. (2014) Environmental sustainability at tourism and recreation areas: A risk assessment-based model. *Journal of Sustainable Tourism* 22(2), 319–338. DOI: 10.1080/09669582.2013.815762.

Ruiz-Ballesteros, E. and Brondizio, E. (2013) Building negotiated agreement: The emergence of community-based tourism in Floreana (Galápagos Islands). *Human Organization* 72(4), 323–335.

Sanchez, P.M. and Adams, K.M. (2008) The Janus-faced character of tourism in Cuba. *Annals of Tourism Research* 35(1), 27–46. DOI:10.1016/j.annals.2007.06.004.

Sharpley, R. and Knight, M. (2009) Tourism and the state in Cuba: From the past to the future. *International Journal of Tourism Research* 11(3), 241–254.

Simon, F.L. (1995) Tourism development in transition economies: The Cuba case. *The Columbia Journal of World Business* 30(1), 26–40.

Simpson, M.C. (2008) Community benefit tourism initiatives: A conceptual oxymoron? *Tourism Management* 29(1), 1–18.

United Nations Environment Program (2017) *Negative impacts of Tourism*. Available at: http://staging.unep.org/resourceefficiency/Business/SectoralActivities/Tourism/FactsandFiguresaboutTourism/ImpactsofTourism/EconomicImpactsofTourism/NegativeEconomicImpactsofTourism/tabid/78784/Default.aspx (accessed 9 October 2017).

Urošević, N. (2014) The Brijuni Islands – recreating paradise: Media representations of an élite Mediterranean resort in the first tourist magazines. *Journal of Tourism History* 6(2–3), 122–138.

Wilkinson, S. (2008) Cuba's tourism 'boom': A curse or a blessing? *Third World Quarterly* 29(5), 979–993.

Williams, A.M. and Baláz, W. (2002) The Czech and Slovak Republics: Conceptual issues in the economic analysis of tourism in transition. *Tourism Management* 23, 37–45.

Winson, A. (2006) Ecotourism and sustainability in Cuba: Does socialism make a difference? *Journal of Sustainable Tourism* 14(1), 6–23. DOI: 10.1080/09669580608668588.

World Travel and Tourism Council (2016) *Travel and Tourism: Economic Impact 2016 Cuba*. WTTC, London. Available at: www.wttc.org/research/economic-research/economic-impact-analysis/country-reports/ (accessed 9 October 2017).

8 Subjective Well-being and Tourism Development in Small Island Destinations

Robertico Croes,* Manuel Rivera and Kelly J. Semrad

University of Central Florida, United States of America

8.1 Introduction

This chapter examines the literature pertaining to the relationship between tourism development and well-being from residents' perspectives. The relationship between tourism development and the well-being of residents impacted by that development is relevant to understand how tourism development unfolds the social dynamics of small islands. It is further relevant given its impact on the productivity and economic prospects of these islands as they are influenced by tourism development. The chapter will address topics that contribute to well-being, such as the relationship between higher incomes and resident happiness, and the role of tourism development in enhancing social comparison as a force in defining resident happiness. Literature gaps as they pertain to subjective well-being and tourism development in small island destinations will be exposed, insights into future research possibilities as they are associated with well-being and tourism development will be offered, and policymaking possibilities that might contribute to the burgeoning relationship between residents' well-being and tourism development will be discussed.

While studies addressing well-being have increased in recent times (Diener, Suh, Lucas and Smith, 1999; Kahneman, Diener and Schwarz, 1999; Keyes, Schmotkin and Ryff, 2002; Seligman, 2011) attempts to define well-being have been ambiguous. This ambiguity has hampered clarity regarding the concept's measurement, and has made the concept's application to human welfare cumbersome. Insofar as mainstream tourism literature is reflecting an increasing trend in evaluative and empirical outputs (Croes, 2016) regarding well-being, then clarity in definition becomes obligatory.

The concept of well-being has been used interchangeably to depict quality of life, happiness and life satisfaction. It covers the spectrum from *hedonia* to *eudaimonia* addressing questions regarding the good life or a life well lived. *Hedonia* references life filled

*E-mail: robertico.croes@ucf.edu

with pleasure, satisfying physical and emotional needs and aiming at the absence of pain. Conversely, *eudaimonia* positions well-being as a meaningful and fulfilled life, anchored in high, life values and personal growth, and aiming at something bigger than oneself. The former refers to feelings or pleasurable experiences while the latter is associated with behaviour or pursuing self-realization (Huta and Ryan, 2010). While wide studies have discussed these traditional views of well-being, it is beyond the scope of this chapter to assess the ethical claims associated with these two concepts. Rather, this chapter departs from a social science perspective and attempts to shed light on the variables that bring bearing to the nature of the relationship between well-being and tourism development.

8.2 The Ontological and Epistemological Debate About Well-being

The debate in the literature regarding the concept of well-being is characterized by its definition, its measurement and its influence on the relationship between tourism development and residents. Traditionally, the concept of well-being has been considered from a utilitarian perspective. Social arrangements were assessed according to whether or not residents' preferences were satisfied. However, making the presumption that preference satisfaction was the informational space with which to assess social arrangements has been met with increasing challenge. An alternative move from the utilitarian perspective to a capabilities perspective (whereupon opportunities available to individuals and the freedom to choose from these opportunities) has emerged. This shift from utilities to capabilities spawned an ontological and epistemological move in how social arrangements are assessed in social sciences.

Although the well-being concept has met with ambiguous perspective in its definition, measurement and influence, the development literature consistently connects the concept with a good life, which includes purpose in life, positive relationships, autonomy and social acceptance. Regarding this, empirical claims respond in general to the question how satisfied are you with your life?

Over time, the concept of well-being has shifted from a one-dimensional conceptualization to a multi-dimensional construct. Initially, well-being was depicted by materialistic conceptions (money buys happiness). Income dominated the evaluation space of well-being; but soon the material primacy (possession and resources) was challenged in the literature, and income has been portrayed as an inadequate conceptualization of well-being. Slowly, well-being equated by income generation shifted towards *how* income was being distributed, resulting in the shift to a new equation of well-being characterized by equality or inequality of resources.

The dimension of inequality as the underpinning of well-being resonated in the basic needs approach and answered questions related to resource distribution. Distribution became relevant in the good life (Croes and Rivera, 2016). However, this conceptualization shift did not placate critics with assertions that well-being included more dimensions than just income and its distribution, such as health and education (Kanbur, 2016). The people upon which well-being was measured may have been vulnerable to their inability to adapt to either increased wealth or deprivation – causing negative effects of material desires and possessions to emerge. This inability could then colour or ignore contributing measurements of intrinsic and extrinsic characteristics that could cause a well-being perspective. A journey from a one-dimensional concept (income) to a multi-dimensional concept of well-being was generated.

As income is a poor metric with which to capture the full array of the diversity of humans in their ability to convert resources into well-being (Sen, 1985), outcomes became the most important unit of analysis to consider when assessing well-being. The main proponent of this shift was Sen in his capability approach as he expanded well-being from income (inputs) to a broader range of outcomes. Because of varied abilities to convert resources into well-being, human diversity and its range of personal traits, wants and needs, and the existing social arrangements are a consideration in defining and measuring well-being. Sen claims that income or utility as measures of well-being placed the wrong focus on what should be assessed in what constitutes the good life. Instead, well-being should be more directly linked to its foundational roots and not to its instrumental antecedents. For Sen a person's well-being can be assessed in their capability to achieve personally and socially valued functionings. At one point, he even equates this with a definition of quality of life.

Sen's capability framework has shifted attention not only towards the impact of outcomes on well-being as functionings or achievements and their connection with social arrangements, but he also expanded the evaluative space of well-being to include opportunities. From Sen's perspective, well-being should also include resource distribution (equality) and the opportunities that an individual requires to lead the life that he or she values. The distinction between achievements (functionings) and opportunities (capabilities) is significant in understanding well-being's requirements in people's conditions to exist, to function and to act on that which they value (Sen, 1999). Nussbaum (2011) and Robeyns (2016) concur that achievements and opportunities are at the core of the capability approach. Viewed from this perspective, the standards defining well-being come to be a range of choices, the freedom to choose according to awareness and aspirations, and the resources to realize these aspirations (Croes, 2012).

The capability approach is anchored on objective life conditions and eschewed life experience as a relevant life domain. Here, the evaluative stance can be objective or subjective. However, for a long time, the objective or subjective have been pitted against each other (Stiglitz, Sen and Fitoussi, 2009). The dichotomy between *hedonia* and *eudaimonia* as perspectives underpinning the concept of well-being addressed, while not sole to well-being assessment, have been revealed in the evaluative stance required to assess well-being. The objective stance purports to view well-being from some objective established conditions, such as income, education and health. These objective conditions are determined by observations outside the subjects. For example, the Human Development Index (HDI), Index of Sustainable Economic Welfare (Daly and Cobb, 1989), the Fordham Index (Miringoff, Miringoff and Opdycke, 2001) and Gross National Happiness (Brooks, 2008) also aim to measure social well-being. However, objective information about human behaviours may not coincide with individuals' own assessment of their situation. Information about how an individual feels or perceives his or her life conditions is neglected because the criteria to assess well-being is imputed or presumed.

Conversely, subjective well-being claims that a person may be in a better position to assess his/her life conditions and opportunities. The subjective stance asserts that the subjects themselves should determine their well-being without outside interference. A person may objectively be satisfied with his/her life even though that life may seem poor or unsatisfying by objective measures. Sometimes people with below-average incomes rank their well-being as high. Therefore, well-being is better understood by asking the individuals directly to evaluate their own situation (Diener and Seligman,

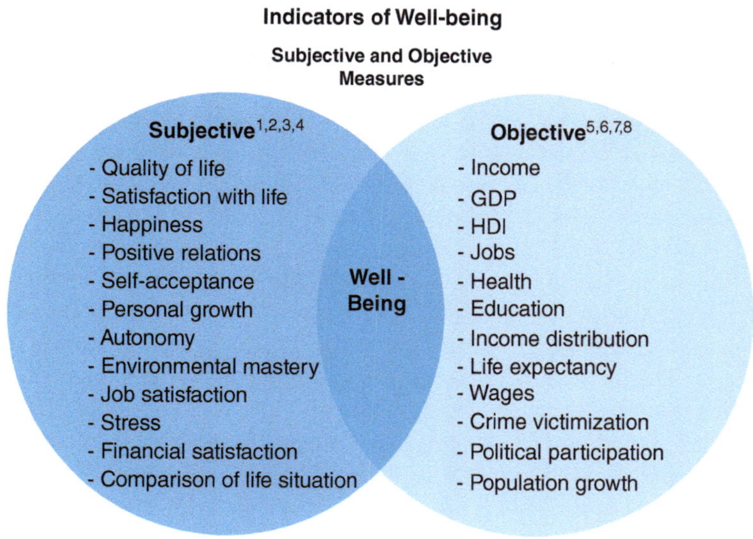

Notes: 1=(Diener *et al.*, 1985); 2=(Diener, 2006); 3=(Kahneman and Krueger, 2006); 4=(Ryan and Deci, 2001); 5=(Bauer, 1966); 6=(Blanchflower and Oswald, 2004); 7=(Easterlin, 2001); 8=(Smith and Clay, 2010)

Fig. 8.1. Measures of well-being. (From authors' own intellectual creation.)

2004). This approach justifies self-assessment in the measurement of subjective happiness when compared with others within a society, and has suggested a new epistemology according to a person's life experience, rather than an experts' constructs. Subjective well-being has been recognized as a key aspect of quality of life, and it becomes crucial to understand how this life experience is being shaped by tourism development. For a comparison of objective and subjective indicators of well-being please see Fig. 8.1.

8.3 Well-being in the Tourism Literature

The tourism literature also reveals the evolving meaning of the well-being construct. Croes (2012) discusses how the three frameworks of the income paradigm, the capability approach and the subjective well-being measurement have been applied in the mainstream tourism literature. Overall, well-being has been used interchangeably with quality of life, happiness and satisfaction with life (Nawijn and Mitas, 2012). The mainstream tourism literature depicts the evolution of well-being as a movement from tourism social impact and residents' attitudes and perceptions regarding tourism development to how individuals feel regarding tourism development. The latter tourism literature feature applied the subjective well-being theory to conceptualize and measure how tourism impacts tourists' well-being.

From Attitudes and Perception to Subjective Well-being

A large group of studies was motivated to use the community as the unit of analysis, including the studies of Andereck *et al.* (2005), Dyer *et al.* (2007), Gursoy *et al.* (2002), McGehee and Andereck (2004) and Wang and Pfister (2008). Other studies, while indicating that their studies focus was the impact of tourism on the well-being of the individual, were essentially still working within the contours of the host community instead of the individual. The unit of analysis of their investigation remained the host community (Yang and Wall, 2009; Diedrich and Garcia-Buades, 2009; McKercher and Ho, 2012; Yang and Li, 2012; Uysal *et al.*, 2012; Woo, Kim and Uysal, 2015). The main feature of this strand is the examination of residents' attitudes, which were considered from a social exchange perspective. This presumed cost–benefit analysis found that tourism was perceived as positive in the socio-economic realm (jobs and income), while the cultural and environmental impacts were perceived as negative. Yet, little is known how each individual resident perceives the impacts of tourism and how tourism impact affects their well-being (Benckendorff, 2009; Uysal *et al.*, 2012).

The focus on attitude and perception regarding the impact of tourism development on the community rather than on how an individual internalizes tourism's impact on his/her well-being does not provide meaningful insights into understanding the relationship between tourism development and well-being. The reason for this flaw is these studies did not focus on how people feel regarding tourism's impact on their life conditions, and how people's perceptions shape variations in their individual well-being (Nawijn and Mitas, 2012; Bimonte and Faralla, 2016). Some studies, such as Anderleck and Nyaupane (2011), focused on well-being items that were more related to residents' way of life than their life conditions or how they felt about their life and tourism in relation to their lives.

Nawijn and Mitas (2012) argue that it was more relevant to investigate how an individual internalizes the impacts of tourism and how this impact would shape the life of the individual. Croes (2016) also contends for the individual as the unit of analysis by claiming that the demand for unique experiences is the driver for destination choice – the result of which is that the inclination for unique experiences thrusts greater tourism impact upon residents. Uniqueness is shaped by the inclusion and interaction with tourists according to residents' everyday life (Croes, Lee and Olson, 2013). That everyday life is bound by multi-factors beyond income and resource distribution.

The lack of studies pertaining to residents' well-being in small island destinations prompted the *Journal of Destination Marketing & Management* to dedicate a special issue on this topic. The special edition reviewed the situation in Aruba in the Caribbean, Mauritius in the Indian Ocean, Fiji in the Pacific, Cyprus in the Mediterranean, Sitka in Alaska and Magnetic Island in Australia (Rivera, Croes and Lee, 2016; Naidoo and Sharpley, 2016; Pratt, McCabe and Movono, 2016; Vogt, Jordan, Grewe and Kruger, 2016; Boukas and Ziakas, 2016; Moscardo and Murphy, 2016). The studies highlighted meaningful well-being aspects, such as the slim relationship between income and well-being, the role of mediating factors, such as social comparison and social cohesion, and inclusion (Croes, 2016).

The shift to the individual as the new unit of analysis prompted a renewed discussion on the definition and meaning of well-being. The reason for this renewed interest in well-being is due mainly to the well-being concept's association with the opportunity to create authentic offerings, enhancing the tourist experience. Croes (2016) articulated this notion:

> The residents' willingness to engage with tourists and the quality of that engagement is founded in the manner in which each resident perceives tourism development as it impinges on his or her well-being. Yet, because well-being cannot be disentangled from a

> person's experience, the same set of objective variables, such as income and jobs, which has been traditionally used to measure well-being, may induce different experiences across different people. (p. 1)

In addition, Nawiin and Mitas (2012) argue that the focus on well-being would spawn higher levels of productivity, and revealing that happy people live longer and have more meaningful relationships.

According to Croes (2012), well-being is a unique construct that integrates three dimensions. The first dimension encompasses the material foundation of life; this places income as the most important determinant of well-being. The second is interrelated with the breadth of opportunities and selections a person must achieve and fulfil those opportunities (e.g. capabilities approach), and to life results. The third dimension echoes the feelings and mindset of an individual regarding life domains. Self-assessment of feelings regarding life domains and sub-domains, such as health, income, job, relationships and inclusion, became the basic dependent variable (Gilbert and Abdullad, 2004; Steyn, Saayman and Nienaber, 2004; and more recently, Corvo, 2011; De Bloom, Guerts and Kompier, 2010; Michalko, Ratz and Bakucz, 2010; Andereck and Nyaupane, 2011; Nawijn and Mitas, 2012; Uysal *et al.*, 2012). The focus of these studies addressing tourists' well-being slowly moved towards residents' well-being. For a comparison of the different determinants of well-being between residents and tourists see Fig. 8.2.

The relationship between tourism and well-being

Residents in small island destinations seem to score high on their life satisfaction scale. Residents in Palma de Mallorca had a mean score of 6.64 on a scale of 10 regarding their life satisfaction, which implies that 66.4% were satisfied with their lives (Nawijn and Mitas, 2012). The score of life satisfaction in Aruba in a 2011

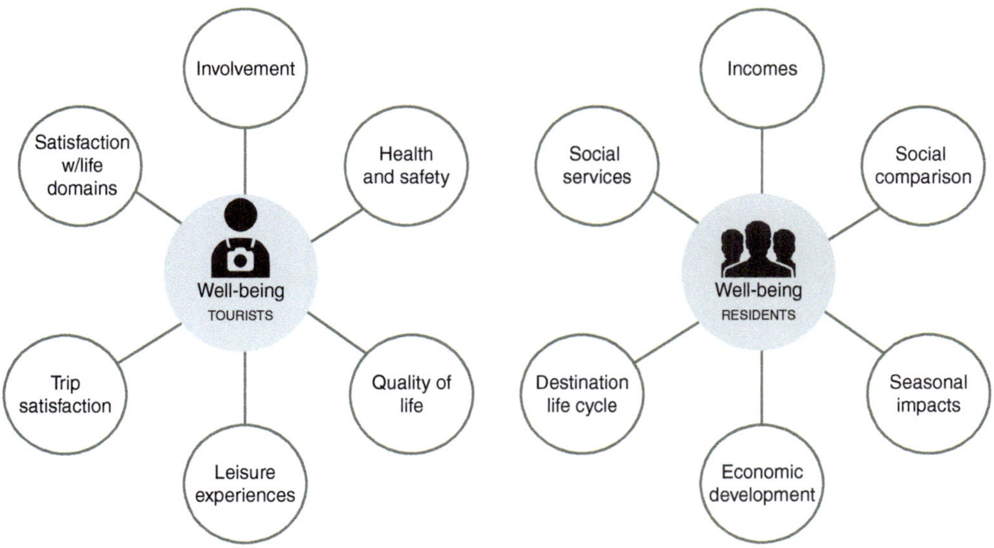

Fig. 8.2. Determinants of well-being. (From authors' own intellectual creation.)

study conducted by Rivera *et al.* (2016) revealed that 76% were satisfied with their lives. A similar question, which was posed in 2016 in Aruba, showed that a similar score (74%) was revealed (Croes *et al.*, 2016). This comparison between 2011 and 2016 studies shows that the score of life satisfaction hovers around 75% and shows a remarkable constant maximum. Ridderstaat *et al.* (2016a) asked a similar question about life satisfaction in 2013 and found that 80.5% of the respondents in Aruba reported that they were satisfied with their lives.

A handful of studies empirically assessed the relationship between tourism development and residents' well-being. These studies were anchored on residents' subjective well-being so the data were drawn from self-reported surveys. Nawijn and Mitas (2012) investigated the relationship between tourism and well-being in Palma de Mallorca. They found that tourism impact was more salient with life satisfaction (the cognitive element of well-being) compared to the affective component, which is happiness. Their study found that tourism impact explained 9.1% of the variance in residents' life satisfaction.

Conversely, Rivera *et al.* (2016) examined the relationship between tourism and well-being. Their study context was the island of Aruba in the Caribbean. Tourism in this study explained only 4% of the variance of residents' well-being induced by tourism. Ridderstaat, Croes and Nijkamp (2016a) also examined this relationship and found that tourism development had a 10% impact on well-being. These three studies follow the findings of Lyubomirksy, Sheldon and Schkade (2005) study, which asserts that subjective well-being research has consistently found that a 10% individual well-being variance is due to life circumstances.

All three studies reveal that tourism development has a statistically significant, positive impact on well-being. These studies also highlight an important consistency with mainstream tourism economics, which is the positive significant relationship between tourism and the local economy. However, Pratt *et al.* (2016) found that with Fiji respondents dependent on tourism self-reported that their life satisfaction level was lower than respondents who were not dependent on tourism. The latter study suggests that the positive relationship between possession and resources and well-being does not always hold with small island destinations. A list of selected Caribbean countries and indicators of tourism performance and well-being is presented in Fig. 8.3.

The Pratt *et al.* (2016) study indicates that the relationship between tourism and well-being does not seem linear or significant. A recent study conducted by Croes *et al.* (2016) in Aruba also found that tourism revealed no significant relationship with well-being. Tourism's influence on well-being seems to diminish. For example, Croes *et al.* (2016) found that tourism played a significant role in shaping well-being in 2011, albeit its impact was modest (only 4%). Bimonte and Faralla (2016) also found that respondents' perceived happiness due to tourism in Follonica, Italy, deteriorated. These results seem consistent with the Easterlin paradox, which references there was not a relationship between well-being (happiness) and income when countries become wealthier (Easterlin, 1974) – although there is considerable debate about whether an Easterlin paradox exists (Deaton, 2008).

The above result prompted an examination of the relationship between some well-being objective conditions as revealed in the Human Development Index (HDI) of some Caribbean islands and their relationship with their economic development level. The latter was depicted by the gross development product (GDP) in PPP dollars.

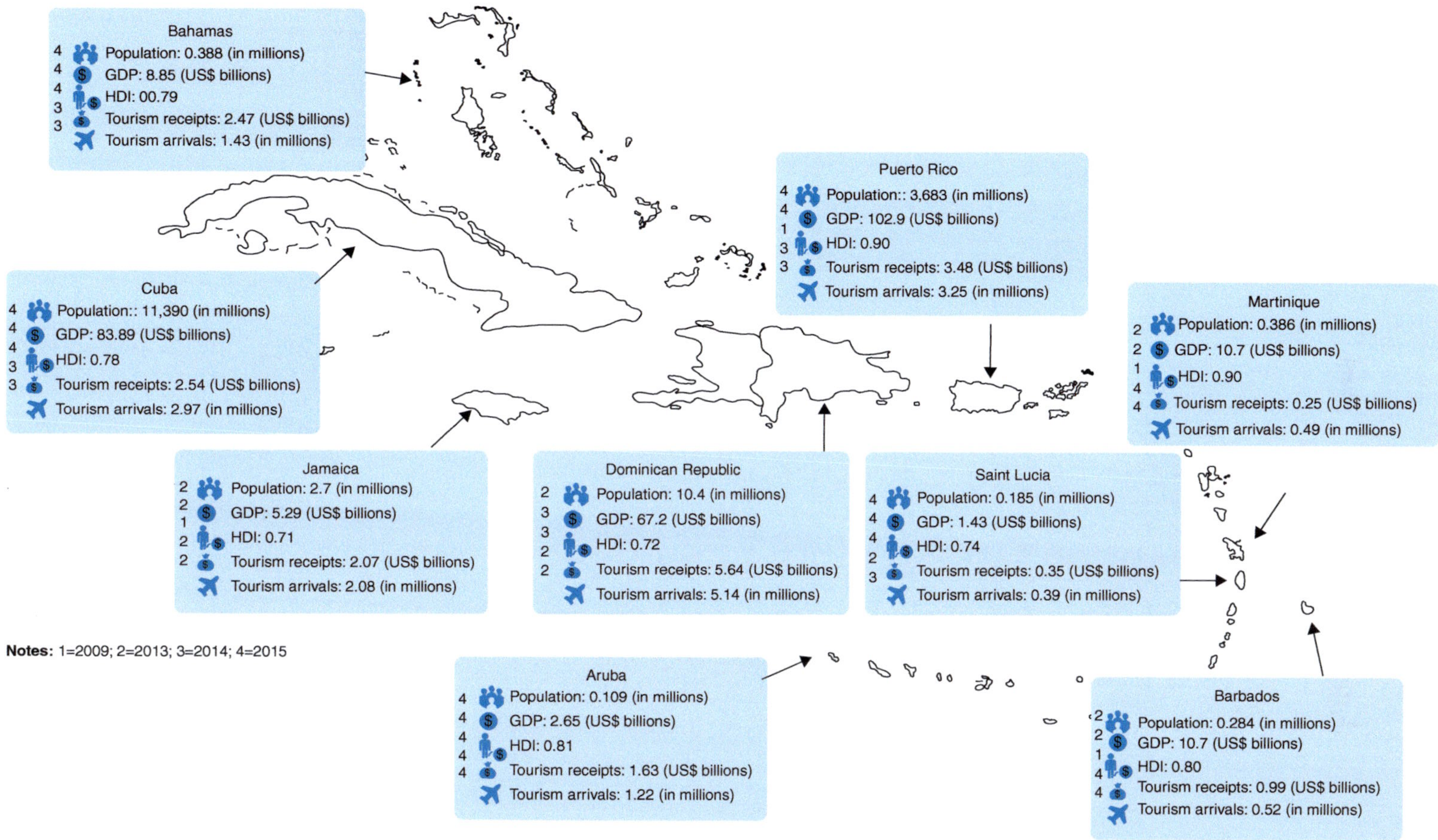

Fig. 8.3. Well-being and economic indicators for selected Caribbean islands. (From authors' own intellectual creation.)

The selected Caribbean islands were Aruba, Barbados, Bahamas, Cuba, Dominica, Dominican Republic, Grenada, Haiti, Jamaica, St. Lucia, St. Kitts and Nevis, St. Vincent and the Grenadines, and Trinidad and Tobago. This selected group of islands reveals variance in tourism development, size, political regime, economic development and poverty. For example, Cuba is the largest island in the Caribbean and is a communist regime. Haiti is the poorest country in the western hemisphere and Barbados, Jamaica and Aruba are mature tourist destinations. Destination maturity refers to the slowing down of arrival growth and bed nights with eventual decline in these rubrics. The destination maturity concept is linked to the product life cycle framework, which suggests that eventually product consumption will spawn negative externalities hindering the repeat visitor process (Butler, 1980; Whitehall and Greenidge, 2000).

The result of the examination is depicted in Fig. 8.4. The results indicate a polynomial relationship, so the relationship between well-being and tourism development becomes weaker as the island becomes wealthier. This result suggests that other factors may be at work mediating the relationship between tourism development and well-being.

Drivers of the relationship between tourism and well-being

Several drivers seem to shape the relationship between tourism development and well-being. For example, Nawijn and Mitas (2012) seem to claim that the link between well-being and tourism development is not contextual, but seems to depend solely on the individual's perception of fulfilment of life under his/her absolute needs and experience. Subjective well-being in this work reveals an association with one's happiness through evaluating one's own past. The relationship between tourism and well-being is not contextual. This claim has been contested by several studies, including Kim, Uysal and Sirgy (2013) and Rivera *et al.* (2016).

Kim *et al.* (2013) used a cross-sectional analysis to examine the relationship between tourism and well-being as defined by life satisfaction. The study found that the strength of the relationship is moderated by the stage of tourism development. Ridderstaat *et al.* (2016a) found that economic development moderates the relationship between tourism and well-being in Aruba. Rivera *et al.* (2016) also found that context matters and shapes the relationship between tourism and well-being. Well-being stems from the interaction of two aspects of a person's life: the fit between an individual's perception of the objective situation and fulfilling needs both from the assessment of one's own situation and one's rank within a group of peers. Well-being is also an outcome of the comparison with others. The study found a positive relationship between tourism development and happiness, and the three mediating variables (income, social comparison and quality of life) considered had an impact of 27% on the variance of the relationship. However, the largest impact came from social comparison, which was five times as large as income. This finding implies that interdependence of preferences takes place, which reveals the relevance of the relative position of individuals for their well-being. The individual compares her own income with that of others and this comparison determines her well-being.

The contention of Rivera *et al.* (2016) that the relationship between tourism development and well-being is mediated was echoed by the study of Bimonte and Faralla (2016). The latter also found that contextual factors matter; the study's result

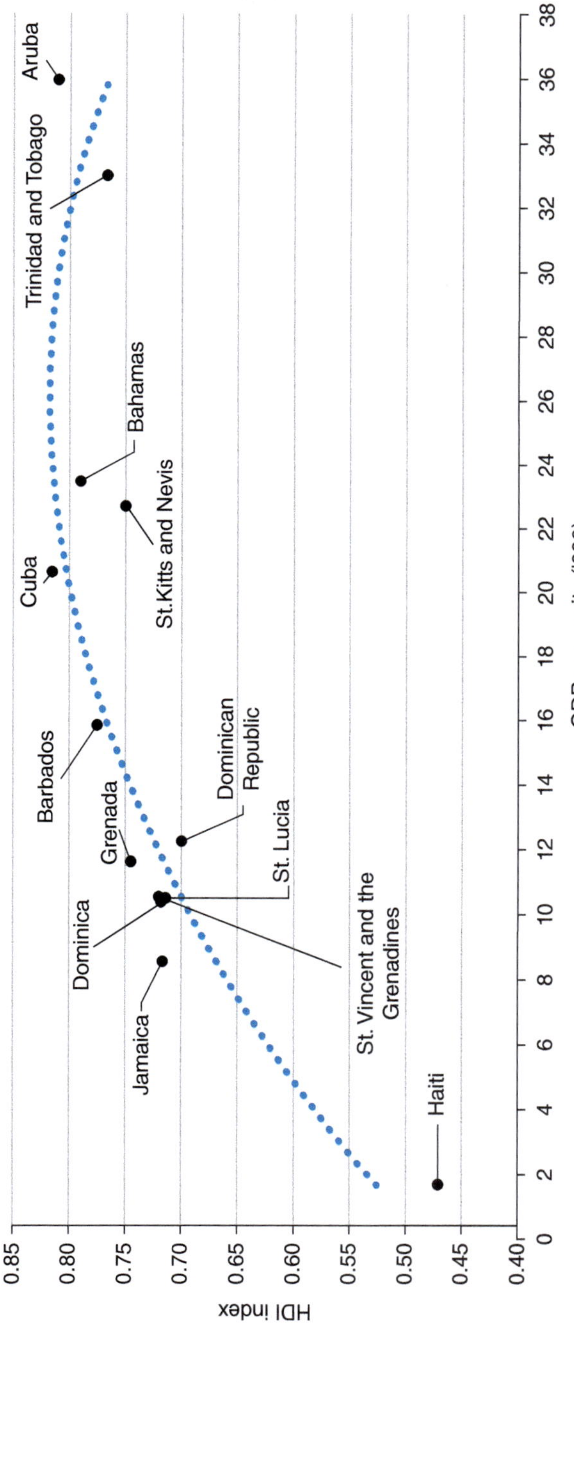

Fig. 8.4. The relationship between HDI and GDP per capita in selected Caribbean countries, 2013. (From authors' own intellectual creation.)

indicates that the tourism season mediates the relationship between tourism and well-being. Croes, Rivera and Semrad (2016) conducted a longitudinal investigation of the relationship between tourism development and well-being (happiness) in Aruba – comparing the impact in 2011 and 2016. The study found that life experiences matter more than personality in shaping people's happiness levels. This result contradicts the set point theory, which asserts that the level of subjective well-being is shaped by heredity and personality traits and remains relatively stable. Happiness does not evolve with time and happiness changes neither for the better nor for the worse regardless of circumstances (Lyubomirksy *et al.*, 2005).

However, Ridderstaat *et al.* (2016b) contend that previous studies examined the one-directional route only between tourism development and well-being as defined by quality of life. Ridderstaat's study investigated this relationship in Aruba applying economic development as a mediating variable. The study stressed important domains that characterize the reciprocal relationship between tourism development and well-being, such as opportunity to get a job, wage or income, amount of sleeping hours, amount of exercise hours and business opportunities. A final interesting aspect in the examination of the impact of tourism development on residents' subjective well-being is the effort to overcome the dichotomous nature of the well-being measurement as revealed in the objective or subjective approaches (Table 8.1).

Table 8.1. Comparison of the findings based on objective and subjective approaches. (From Ridderstaat, 2015, adapted with permission.)

Objective approach	Hypothesis	Subjective approach
✔ Accepted	Tourism development has a direct impact on the quality of life of residents	✔ Accepted
✔ Accepted	Quality of life of residents has a direct impact on the tourism development of a destination	✖ Rejected
✔ Accepted	Tourism development has a direct impact on long-run economic development	✔ Accepted
✔ Accepted	Economic development at a destination has a direct impact on tourism development	✔ Accepted
✖ Rejected	Economic development at a destination has a direct impact on residents' quality of life	✔ Accepted
✔ Accepted	Quality of life of residents has a direct impact on the economic development of a destination	✔ Accepted
✖ Rejected	Tourism development indirectly impacts the quality of life of residents through the mediating role of economic development	✔ Accepted
✔ Accepted	Quality of life of residents indirectly impacts tourism development through the mediating role of economic development	✔ Accepted

Following Diener *et al.* (1999) and Stiglitz *et al.* (2009) who suggested to combine objective and subjective measurements of well-being, Ridderstaat *et al.* (2016a, b) examined the relationship between well-being and tourism development from an objective and a subjective perspective. The two studies examined the same hypotheses, allowing for a comparison of the findings of both investigations. Generally, the results validated most of the applied hypotheses, and only in three cases did the findings indicate different results (see Table 8.1). These outcomes validate to a large extent the overlap between the objective and subjective approaches towards the relationship between tourism development and residents' well-being with Aruba. The findings also reveal that material conditions, as expressed by economic development, do not seem to overlap between objective and subjective well-being when considering the relationship between tourism and well-being. Table 8.1 shows the results of evaluating the hypotheses.

8.4 Case Study: Aruba

This case study focuses on tourism development in relation to residents' well-being in Aruba. Since 1986, Aruba has grown steadily as an attractive tourist destination, and with experienced tourism development on the island for the last 50 years. Aruba's main tourism products are sun, sand and sea; and Aruba also distinguishes itself among other Caribbean islands for having friendly local people, accepting US dollars on the island, providing a clean, safe water supply and the general safety of the island. Aruba has gained its popularity and loyalty among tourists because Aruba has experienced a strong repeat visitation rate of approximately 60% with strong levels of satisfaction by the tourists (Central Bureau of Statistics, 2011). Tourism has brought great economic benefits to Aruba broadly shared by many (Croes, 2011). Compared to other Caribbean destinations, Aruba provides a good case study to assess the impact of tourism on the residents' satisfaction with life.

The research model used in this study is presented in Fig. 8.5. For small island destinations that are tourism dependent, it is essential to pursue the well-being of its residents. If tourism could impact people's lives and if residents understand the potential of tourism to be positive, then a perception of tourism development should indicate people to be more satisfied with their lives. Therefore, it was proposed that tourism development could trigger a state of satisfaction with life in individuals.

The impact of tourism on a small island destination may also affect perceptions of social rank among residents. These concepts of social comparison are deeply associated with an individual's own culture and life history. The subjective assessments of how individuals respond to external conditions in their lives appear important in small islands. This is mainly because external conditions, such as the benefits of tourism development, affect individuals' social ranking (Easterlin, 1974; Felce and Perry, 1995; Suls and Wheeler, 2000; Frank, 2008). Therefore, bygone experiences, judgement and comparison to others can sway and influence the residents' assessment of the external conditions and of their life situations as they are influenced by those external conditions. We hypothesize that comparisons of life situations in Aruba mediate the relationship between tourism development and the residents' satisfaction with life.

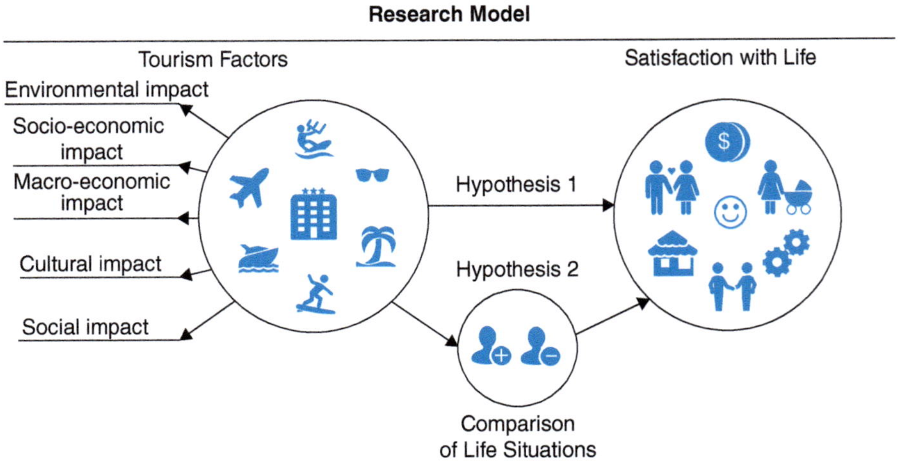

Proposed Hypothesis

Hypothesis 1	Tourism development has a direct impact on residents' satisfaction with personal life
Hypothesis 2	Tourism development has a direct and indirect impact on residents' satisfaction with personal life through comparison of life situation

Fig. 8.5. Proposed research model and hypothesis. (From authors' own intellectual creation.)

Research model: Tourism development and residents' well-being

With Aruba, tourism development is defined as a five-construct measurement model. The construct included social impact, cultural impact, environmental impact, socio-economic impact and macro-economic impact (Liu and Var, 1986; Ap and Crompton, 1988; Lankford and Howard, 1994; Weaver and Lawton, 2001; Kim, 2002; Tosun, 2002). In addition, to test hypothesis 2, the comparative position of a respondent to the total population was gauged by asking respondents to compare his/her life to 'most other people', and to indicate how much better or worse his/her life is (Andrews and Withey, 1976). The satisfaction with personal life (SWPL) scale was measured consisting of four items – which serves as an endogenous construct (e.g. In most ways my life is close to being ideal).

The statistical method used to test the proposed hypothesis is structural equation modelling. The statistical procedures to analyse the data include applying the maximum likelihood estimation method based on its goodness-of-fit indices, reliability and the convergent and discriminant validity of the constructs. The observed variables were loaded on their specified latent construct and indicate the internal consistency of items. The composite reliabilities for the five-construct measurement model ranged from 0.75 to 0.84, exceeding the recommended threshold of 0.7 (Bagozzi and Yi, 1988).

The results of a variance extracted test showed that the average variance extracted (AVE) values were above the suggested cut-off of 0.50, and the average shared squared variance (ASV) of each construct was smaller than the AVE to meet qualifications of convergent and discriminant validity (Fornell and Larcker, 1981; Hair, Black, Barry and Anderson, 2010). The overall goodness-of-fit indices showed that the five-construct model had an acceptable fit with the data (X^2 = 431.484, df = 142, p < 0.001, RMSEA = .049, CFI = .957, NFI = .938).

Sample profile

1,200 surveys were distributed and 925 surveys were collected (response rate of 77.1%). After deleting incomplete and unusable surveys, 915 surveys were used for the data analysis, which represents 1.16% of the adult population in Aruba. The demographic profile of the sample is presented in Fig. 8.6.

Structural equation model

A second-order latent factor of tourism development consisted of five first-order constructs: social impact, cultural impact, environmental impact, socio-economic impact and macro-economic impact. Tourism development, a second-order latent factor, accounts for the covariance between constructs, while first-order factors account for covariance between the observed variables (Hair *et al.*, 2010).

Each of the first-order constructs contributed to the account of the second-order construct of tourism development: environmental impact (γ = 0.90, p < 0.001), socio-economic impact (γ = 0.87, p < 0.001), macro-economic impact (γ = 0.83, p < 0.001), cultural impact (γ = 0.74, p < 0.001) and social impact (γ = 22, p < 0.001). Figs 8.7 and 8.8 present the results from the structural model. The first hypothesis expected that tourism development positively would lead to residents' satisfaction with personal life. The results showed that tourism development has a positive and strong direct relationship with residents' satisfaction with personal life (γ = 0.28, p < 0.001). The fitness of the model was good (χ^2 (165) = 663.076, p < 0.001, CFI = 0.932, NFI = 0.912; RMSEA = 0.057; *** p < 0.001). Hence, this hypothesis was supported indicating that the residents are more satisfied with their lives as they see positive influences of tourism development on a destination. Tourism development dimensions explained 8% of the residents' satisfaction with their personal lives.

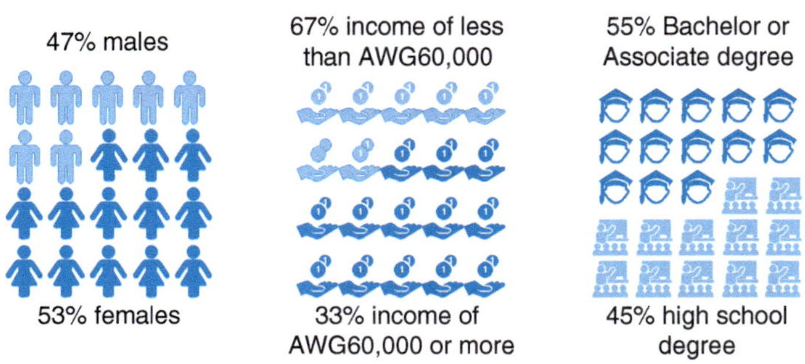

Fig. 8.6. Sample profile and demographics. (From authors' own intellectual creation.)

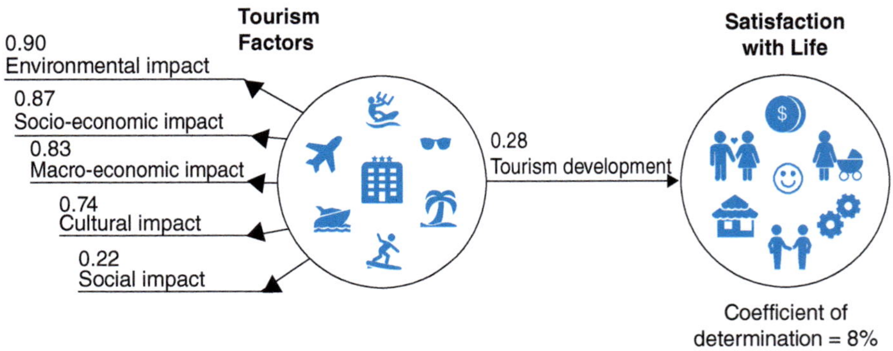

Fig. 8.7. Results for hypothesis 1. (From authors' own intellectual creation.)

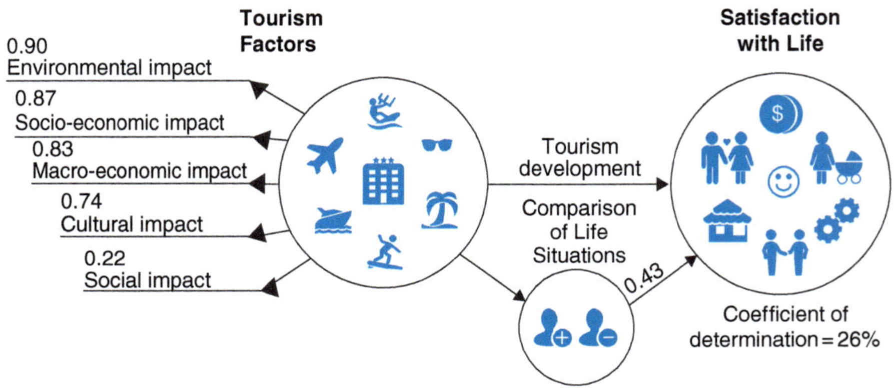

Fig. 8.8. Results hypothesis 2. (From authors' own intellectual creation.)

The second hypothesis predicted that tourism development would be both directly and indirectly related to residents' satisfaction with personal life through comparison of life situation. The fitness of the structural model was good (χ^2 (183) = 709.923, p < 0.001, CFI = 0.931, NFI = 0.909; RMSEA = 0.056; *** p < 0.001). The results indicated that a direct effect between tourism development and satisfaction with personal life was significant (γ = 0.23, p < 0.001). In addition, there were indirect effects between tourism development and comparison life situation (γ = 0.12, p < 0.001) and between comparison of life situation and satisfaction with personal life (γ = 0.43, p < 0.001) (see Fig. 8.8). Comparison of life situation played a significant role in the relationship between tourism development and residents' well-being. Considering comparison of life situation, tourism development explained 26% of the residents' well-being.

8.5 Discussion and Future Research

The chapter reveals that the examination between tourism and residents' well-being is still in its infancy. This situation is salient with warm-water island destinations.

Notwithstanding, some meaningful insights emerged regarding the determinants shaping the relationship between tourism and residents' well-being. Tourism may deliver more incomes to residents. However, as this chapter indicates, raising incomes do not necessarily ensure greater well-being. The connection between income and residents' well-being is not self-evident and is contingent by many intervening factors, some which are beyond the control of the resident. These factors involve, for example, age (the needs of the young versus old person), gender and social roles (women and family) and location (drought and crime).

If the connectedness between income and well-being is not self-evident, perhaps higher income should not be the final goal of tourism development. Applying the connectedness assumption to policy could lead to perverse effects potentially reducing residents' well-being. Relative income seems more important than absolute income in determining residents' well-being, so social comparison is the mechanism for how residents will assess their income in relation to a comparison group (the average of the society, people of similar socio-economic characteristics, neighbours, family, etc.). The logic of this mechanism is that residents are happy if they have more compared to others or unhappy otherwise. The policy implication is that public policy intervention should be focused more on distributive effects of tourism rather than raising incomes through tourism. Residents could alter their expectations with rising incomes rendering only small or ephemeral effects of higher income on well-being. Determining whether the connectedness between tourism (income) and well-being is moderated or mediated by social comparison (distributive effects) or adaptation (comparing past income with present income) warrant important policy considerations with consequential policy implications and outcomes.

However, to make those policy considerations, two foundational issues emerge associated with the relationship between tourism and well-being. The first issue pertains to the relevance of pursuing well-being or happiness as the ultimate policy goal. The premise of residents' well-being studies is that satisfaction with life and happiness matter for policy. Is the pursuit of well-being or happiness a meaningful policy objective? What is the weight of well-being or happiness compared to growth and fiscal stability? What about inter-temporal considerations? Growth may make individuals unhappy in the short-term, because they may find their existence frustrating or growth may make them feel in a treadmill because higher incomes lead only to a short-lived or small increase in well-being. However, eventually growth may guarantee more well-being. Individuals may fail to make the right choices that might not be in their or society's long-term interest. Psychologists have warned against erratic behaviour spawned by irrelevant factors of choice. This circumstance is related to the second issue that concerns the role of public policy. Should public policy be directed by enriching people's capabilities set, by preference satisfaction, or by subjective well-being?

How well-being policies are designed, measured and monitored is contingent on residents' freedom to function, to seize opportunities, to make choices and to take actions based on free will. Sen (1999) characterizes actions based on free will as agency. The set of opportunities does not depend only on social arrangements; opportunities also depend on agency and how agency interacts with these opportunities. However, the nature of the intervention of agency on the relationship between tourism and residents' well-being is not clear. Is the nature progressive? Does a sequential causal path exist from acquisition and possession of resources to well-being and the good life? Does this causal path include a pattern from agency acquisition and possession of resources through tourism to conversion into functionings (achievements) that are meaningful and valued capabilities, which lead to well-being and the good life? Or are there other paths possible; and what would the configuration of these paths be? Are these paths, for example, recurrent or non-recurrent?

Measuring and monitoring the effectiveness of well-being policies depend ultimately on the definition of the informational space to assess residents' well-being. How do we measure the impact of tourism development on residents' well-being? The objective and subjective perspectives do not differ only by their contents, but also by their epistemology. The objective perspective is viewed through the lens of the observer who assesses the impact through some criteria determined by a third party outside of the subject. The subjective perspective is defined directly by the subject through his/her expression of feelings, emotions and cognitive appreciation about his/her life circumstances. Can these two perspectives be reconciled? Can meaningful and valued capabilities be assessed only subjectively without compromising or clouding the subjective judgement stemming from the individual's coping strategies with reality? Or, is it possible to come up with a list of objective criteria that facilitates assessment without patronizing the individual and without compromising the interaction of individual expectation, aspiration and reality?

These are questions – the answers of which may assist in our quest for theory building, which in the process may make meaningful contributions to the policy-making that might enhance well-being. These and other questions should be addressed in future research.

References

Andereck, K.L. and Nyaupane, G.P. (2011) Exploring the nature of tourism and quality of life perceptions among residents. *Journal of Travel Research* 50(3), 248–260.

Andereck, K.L., Valentine, K.M., Knopf, R.C. and Vogt, C.A. (2005) Residents' perceptions of community tourism impacts. *Annals of Tourism Research* 32(4), 1056–1076.

Andereck, K., Valentine, K., Vogt, C. and Knopf, R. (2007) A cross-cultural analysis of tourism and quality of life perceptions. *Journal of Sustainable Tourism* 15(5), 483–502.

Andrews, F.M. and Withey, S.B. (1976) Measuring global well-being. In: *Social Indicators of Well-Being, Americans' Perception of Quality of Life*. Plenum Press, New York, pp. 63–106.

Ap, J. and Crompton, J. (1998) Developing and testing tourism impact scale. *Journal of Travel Research* 37(2), 120–130.

Bagozzi, R.P. and Yi, Y. (1988) On the evaluation of structural equation models. *Journal of the Academy of Marketing Science* 16(1), 74–94.

Benckendorff, P., Edwards, D., Jurowski, C., Liburd, J.J., Miller, G. and Moscardo, G. (2009) Exploring the future of tourism and quality of life. *Tourism and Hospitality Research* 9(2), 171–183.

Bimonte, S. and Faralla, V. (2016) Does residents' perceived life satisfaction vary with tourist season? A two-step survey in a Mediterranean destination. *Tourism Management* 55, 198–208.

Brooks, A. (2008) *Gross National Happiness: Why Happiness Matters for America and How We Can Get More of It*. Basic Books, New York.

Boukas and Ziakas, (2016) Tourism policy and residents' well-being in Cyprus: Opportunities and challenges for developing an inside-out destination management approach. *Journal of Destination Marketing & Management* 5(1), 44–54.

Butler, R.W. (1980) The concept of a tourist area cycle of evolution: implications for management of resources. *The Canadian Geographer* 24(1), 5–16.

Central Bureau of Statistics (2011) 2011 Tourism Reports. Aruba. Available at: http://cbs.aw/wp/ (accessed 9 October 2017).

Corvo, P. (2011) The pursuit of happiness and the globalized tourist. *Social Indicators Research* 102(1), 93–97.

Croes, R. (2012) Assessing tourism development from Sen's capability approach. *Journal of Travel Research* 51(5), 542–554.

Croes, R. (2016) Connecting tourism development with small island destinations and with the well-being of the island residents. *Journal of Destination Marketing & Management* 5, 1–4.

Croes, R. and Rivera, M. (2016) *Poverty Alleviation through Tourism Development: A Comprehensive and Integrated Approach*. Apple Academic Press, Waretown, New Jersey.

Croes, R., Lee, S. and Olson, E. (2013) Authenticity in tourism in small island destinations: the residents' perspective. *Journal of Tourism and Cultural Change* 11(1–2), 1–20.

Croes, R., Rivera, M. and Semrad, K. (2016) *Happiness and tourism in Aruba. Insights from the 2016 happiness survey*. Dick Pope Sr. Institute for Tourism Studies, Orlando, Florida.

Daly, H. and Cobb, J. (1989) *For the Common Good*. Beacon Press, Boston, Massachusetts.

De Bloom, J., Guerts, S.A.E. and Kompier, M.A.J. (2012) Vacation (after-) effects on employee health and well-being, and the role of vacation activities, experiences and sleep. *Journal of Happiness Studies* 14(2), 613–633.

Deaton, A. (2008) Income, health, and well-being around the world: evidence from the Gallup World Poll. *Journal of Economic Perspectives* 22(2), 53–72.

Diedrich, A. and García-Buades, E. (2009) Local perceptions of tourism as indicators of destination decline. *Tourism Management* 30(4), 512–521.

Diener, E. and Seligman, M. (2004) Beyond money toward an economy of well-being. *Psychological Science in the Public Interest* 5(1), 1–31.

Diener, E., Suh, M., Lucas, E. and Smith, H. (1999) Subjective well-being: three decades of progress. *Psychological Bulletin* 125(2), 276–302.

Dyer, P., Gursoy, D., Sharma, B. and Cater, J. (2007) Structural modeling of resident perceptions of tourism and associated development on the Sunshine Coast, Australia. *Tourism Management* 28, 409–422.

Easterlin, R. (1974) Does economic growth enhance the human lot? Some empirical evidence. In: David, P. and Reder, M. (eds) *Nations and Households in Economic Growth: Essays in Honor of Moses Abramovitz*. Apple Academic, New York, pp. 89–125.

Felce, D. and Perry, J. (1995) Quality of life: its definition and measurement. *Research in Developmental Disabilities* 16(1), 51–74.

Fornell, C. and Larcker, D.F. (1981) Structural equation models with unobservable variables and measurement error: algebra and statistics. *Journal of Marketing Research*, 382–388.

Frank, R.H. (2008) Should public policy respond to positional externalities? *Journal of Public Economics* 92(8–9), 1777–1786.

Gilbert, D. and Abdullah, J. (2004) Holiday taking and the sense of well-being. *Annals of Tourism Research* 31(1), 103–121.

Gursoy, D., Jurowski, C. and Uysal, M. (2002) Resident attitudes: a structural modeling approach. *Annals of Tourism Research* 29(1), 79–105.

Hair J.F., Black, W.C., Barry, J.B. and Anderson, R.E. (2010) *Multivariate Data Analysis*. Prentice Hall, Upple Saddle river, New Jersey.

Huta, V. and Ryan, R.M. (2010) Pursuing pleasure or virtue: the differential and overlapping well-being benefits of hedonic and eudaimonic motives. *Journal of Happiness Studies* 11, 735–762.

Kahneman, D., Diener, E., and Schwarz, N. (eds) (1999) *Well-being: Foundations of Hedonic Psychology*. Russell Sage Foundation Press, New York.

Kanbur, R. (2016) Capability, Opportunity, Outcome and Equality. Available at: http://ageconsearch.umn.edu/record/250027?ln=en (acessed 8 April 2017).

Keyes, C.L.M., Shmotkin, D. and Ryff, C.D. (2002) Optimizing well-being: the empirical encounter of two traditions. *Journal of Personality and Social Psychology* 82, 1007–1022.

Kim, K. (2002) The effects of tourism impacts upon quality of life of residents in the community. PhD dissertation, Virginia Polytechnic Institute and State University, Blacksburg, Virginia.

Kim, K., Uysal, M. and Sirgy, M. (2013) How does tourism in a community impact the quality of life of community residents? *Tourism Management* 36, 527–540.

Lankford, S.V. and Howard, D.R. (1994) Developing a tourism impacts attitude scale. *Annals of Tourism Research* 21(1), 121–139.

Liu, J.C. and Var, T. (1986) Resident attitudes toward tourism impacts in Hawaii. *Annals of Tourism Research* 13, 193–214.

Lyubomirksy, S., Sheldon, K. and Schkade, D. (2005) Pursuing happiness: the architecture of sustainable change. *Review of General Psychology* 9(2), 111–131.

McGehee, N. and Andereck, K. (2004) Factors predicting rural residents' support of tourism. *Journal of Travel Research* 43, 131–140.

McKercher, B. and Ho, P. (2012) Cultural tourism and the enhancement of quality of life. In: Uysal, M., Perdue, R. and Sirgy, M. (eds) *Handbook of Tourism and Quality of Life Research: Enhancing the Lives of Tourists and Residents of Host Communities*. Springer, London, pp. 341–358.

Michalko, G., Ratz, T. and Bakucz, M. (2010) A theoretical journey along the borders between welfare and well-being: economic aspects based on observations in Hungary. *Megatrend Review* 7(1), 231–248.

Miringoff, M., Miringoff, M. and Opdycke, S. (2001) *The Social Report: A Deeper View of Prosperity: Assessing the Progress of America by Monitoring the Well-being of its People*. Fordham Institute for Innovation in Social Policy, Tarrytown, New York.

Moscardo, G. and Murphy, L. (2016) Using destination community wellbeing to assess tourist markets: a case study of Magnetic Island, Australia. *Journal of Destination Marketing & Management* 5(1), 55–64.

Naidoo, P. and Sharpley, R. (2016) Local perceptions of the relative contributions of enclave tourism and agritourism to community well-being: the case of Mauritius. *Journal of Destination Marketing & Management* 5(1), 16–25.

Nawijn, J. and Mitas, O. (2012) Resident attitudes to tourism and their effect on subjective well-being. The case of Palma de Mallorca. *Journal of Travel Research* 51(5), 531–541.

Nussbaum, M. (2011) *Creating capabilities: the human development approach*. The Belknap Press of Harvard University Press, Cambridge, Massachusetts.

Pratt, S., McCabe, S. and Movono, A. (2016) Gross happiness of a 'tourism' village in Fiji. *Journal of Destination Marketing & Management* 5(1), 26–35.

Ridderstaat, J. (2015) Studies on determinants of tourism demand dynamics in a small island destination: the case of Aruba. Ph.D thesis, Free University, Amsterdam, Netherlands.

Ridderstaat, J., Croes, R. and Nijkamp, P. (2016a) A two-way causal chain between tourism development and quality of life in a small island destination: an empirical analysis. *Journal of Sustainable Tourism* 24(10), 1461–1479.

Ridderstaat, J., Croes, R. and Nijkamp, P. (2016b) The tourism development-quality of life nexus in a small island destination. *Journal of Travel Research* 55(1), 79–94.

Rivera, M., Croes, R. and Lee, S. (2016) Tourism development and happiness: a residents' perspective. *Journal of Destination Marketing & Management* 5(1), 5–15.

Robeyns, I. (2016) Capabilitarianism. *Journal of Human Development and Capabilities* 17(3), 397–414.

Seligman, M.E.P. (2011) *Flourish – A New Understanding of Happiness and Well-being – and How to Achieve Them*. Nicholas Brealey Publishing, London.

Sen, A. (1985) *Commodities and Capabilities*. Elsevier Science Publisher, Oxford, UK.

Sen, A. (1999) *Development as Freedom*. Anchor Books, New York.

Steyn, S., Saayman, M. and Nienaber, A. (2004) The impact of tourist and travel activities on facets of psychological well-being. *South African Journal for Research in Sport, Physical Education and Recreation* 26(1), 97–106

Stiglitz, J., Sen, A. and Fitoussi, J.P. (2009) *Report by the Commission on the Measurement of Economic Performance and Social Progress.* Community Indicators Consortium, Issaquah, Washington.

Suls, J. and Wheeler, L. (2000) *Handbook of Social Comparison: Theory and Research.* Kluwer Academic/ Plenum Publsihers, New York.

Tosun, C. (2002) Host perceptions of impacts: A comparative tourism study. *Annals of Tourism Research* 29 (1), 231–253.

Uysal, M., Perdue, R. and Sirgy, M (2012) (eds) *Handbook of Tourism and Quality-of-Life Research: Enhancing the Lives of Tourists and Residents of Host Communities.* Springer, London.

Vogt, C., Jordan, E., Grewe, N. and Kruger, L. (2016) Collaborative tourism planning and subjective well-being in a small island destination. *Journal of Destination Marketing & Management* 5(1), 36–43.

Wang, Y. and Pfister, R. (2008) Residents' attitudes toward tourism and perceived personal benefits in a rural community. *Journal of Travel Research* 47, 84–93.

Weaver, D. and Lawton, L. (2001) Resident perceptions in the urban-rural fringe. *Annals of Tourism Research* 28(2), 349–458.

Whitehall, P. and Greenidge, K. (2000) Tourism maturity and demand. *Caribbean Centre for Monetary Studies* 5 & 6, 161–189.

Woo, E., Kim, H. and Uysal, M. (2015) Life satisfaction and support for tourism development. *Annals of Tourism Research* 50, 84–97.

Yang, L. and Li, X. (2012) Ethnic tourism and resident quality of life. In: Uysal, M., Perdue, R. and Sirgy, M. (eds) *Handbook of Tourism and Quality of Life Research: Enhancing the Lives of Tourists and Residents of Host Communities.* Springer, London, 373–387.

Yang, L. and Wall, G. (2009) Ethnic tourism: a framework and an application. *Tourism Management* 30(4), 559–570.

PART III: Island Tourism Marketing and Management

9 Tourism in the Seychelles: Trends and Experiences

Neelu Seetaram[1*] and Brigitte Joubert[2]

[1]Bournemouth University, United Kingdom; [2]Seychelles Tourism Academy, Seychelles

9.1 Introduction

Research suggests that the most successful destinations are the ones that can identify and adapt to changes in the tourism industry; particularly to the interests and perception of tourists across varying demographics (WEF, 2015). This is why markets research and studies on consumer behaviour in tourism are among the most prominent areas of tourism research. Such studies are facilitated by the recent development of the Web 2.0, which has enabled online communications among consumers and producers. It has led to increasing consumer reliance on travel reviews in the planning stage of their trips (Buhalis and Law, 2008) as personal recommendations from a wide number of travellers with first-hand experience of the product or destination is preferred in order to reap benefits and obtain the highest level of consumer satisfaction through a more rewarding experience (Filieri and McLeay, 2014).

Seychelles is a small island developing state in the Indian Ocean, which relies heavily on tourism as a source of income and employment. Seychelles has been highly successful in developing the tourism sector over the last four decades and this has been largely enabled by the government's commitment and effectiveness at implementing related policies. The state is now at the position where it not only needs to maintain its position as a world-class destination, but it also faces stronger competition from existing competitors and emerging destinations. In order to remain competitive, the islands are re-thinking their policies and strategies. However, according to Croes (2006, p. 454) 'basing any strategy on a supply perspective is not conducive to keeping these small economies competitive in the era of globalisation' and, therefore, as the global conditions change, it is postulated that future policies and investment aiming at maintaining the high-quality experience that the destination offers will be more effective, by taking into account the opinions of the consumers.

*E-mail: nseetaram@boumemouth.ac.uk

The purpose of this study is twofold. First, it analyses the nature of the tourism demand in Seychelles. Second, it uses the netnographic approach to collect and analyse qualitative data from the discussion forum on Seychelles on TripAdvisor in order to understand the consumers' perspective and their experience of an archipelago as a tourism destination. The netnograhic approach allows for the analysis of consumer behaviour on the Internet to provide useful insights on consumers' perceptions of Seychelles as a destination. It can have advantages over ethnography by enabling the researcher to remain invisible and also trace back online community information from the past (Kozinets, 2010). The data collected are compared to the qualitative data collected by National Bureau of Statistics (NBS) through an exit survey of tourists.

9.2 The Seychelles Islands

The Republic of Seychelles is located in the western Indian Ocean, east of the African continent. The archipelago is situated south of the equator and spans 1.4 million square kilometres. It is made up of 115 islands. The total land area is 177 square miles. The capital, Victoria, is situated on the largest island, Mahé, which is 17 miles long and 7 miles at its widest point. The temperature in the capital ranges from 24°C to 30°C and varies little over the year. The Seychelles is a democracy and has enjoyed political stability over the last 40 years. The population is approximately 92,000 (NBS, 2015a). Tourism takes place on several of the islands. However, the islands of Mahé, Praslin and La Digue, the main inhabited islands, supply the highest number of beds (STB, 2012). As from 2014, the country has been promoted from the upper-middle income category countries to high income by the World Bank (2017). Tourism and fishing (STB, 2012) are the main sources of income on the islands. The tourism capital of the islands may be examined using Couch and Ritchie's (1999) concept of core resources and attractors. This is compiled in Table 9.1.

Tourism policy in Seychelles

The Seychelles Tourism Board (STB) is a semi-autonomous body responsible for marketing of the destination. The Act that governs the actions of the STB is currently under review. The Seychelles Tourism Master Plan (STB, 2012) is the policy document on tourism in the archipelago. It replaced the 'Vision 21 – Tourism Development in Seychelles 2001–2010'.

Economic contribution of tourism in Seychelles

Tourism is by far the main contributor to Seychelles' economy (Laporte, 2013). The World Travel & Tourism Council (WTTC, 2015) report reveals a total of $0.8 billion was collected in terms of tourism-related earnings. Although this is below the $19.4 billion world average, it remains highly significant for the archipelago. The overall contribution to employment and GDP were 56.0% and 56.7%, respectively, in 2014.

Table 9.1. Core resource and attractors of Seychelles. (From data compiled by the authors.)

Category	Examples
Physiography and climate	Warm temperature throughout the year ideal for the nature-based activities on offer More than 50% of Seychelles land is protected under law Tourists can choose among more than 60 beaches on Mahé La Digue island offers an award-winning beach The archipelago is located outside the cyclonic belt of the Indian Ocean, which is an advantage over its competitors, e.g. Mauritius The island is home to a vast number of endemic species of fauna and flora. A few of the most popular are the giant tortoise and the coco de mer and the jelly fish tree
Culture and history	Local dance has African (e.g. 'Moutya') and European ('Kamtole' and 'Contredance') origins The local dishes draw influence from the three continents Africa, Europe and Asia Religious architecture includes churches and temples Numerous buildings of the colonial times with historical importance to Seychelles and the Indian Ocean.
Mix of activities	Activities are predominantly nature-based and include fishing, trails, hiking and snorkelling
Special events	Eco-friendly marathon International carnival
Entertainment	Open air food stalls in town and the beach on designated days Local dance clubs
Superstructure	Modern airport home to the local carrier Air Seychelles Transport systems linking the islands by ferry and by air Hotels and resorts of international standards Provision of budget accommodation by smaller local businesses.
Market ties	Strong cultural and historical links to the three continents that the ancestors of the inhabitants of Seychelles originated from (Europe, Asia and Africa)

Consequently, these figures reveal a very heavy dependence on this sector and it is not surprising that Seychelles is ranked seventh out of 138 countries in 2014 (WTTC, 2015) for its reliance on its tourism sector. The only other nation in the Indian Ocean that is ranked higher than Seychelles is Maldives at the fifth position. This can be a major source of vulnerability for the country. Table 9.3 compares the size of the tourism industry in Seychelles to those of its two main competitors. The size of the tourism industry in Seychelles is not as big as in Maldives, the islands are, however, more reliant on tourism than Mauritius.

The relatively high dependence on the tourism industry is linked to the government's decision to focus on this sector due to limited suitable land for agriculture (Gabbay and Ghosh, 2003). The small population of the country also rules out the possibility of diversifying in other sectors as the opportunity to develop competitive advantage in manufacturing or the finance sector is limited in the absence of a readily available trained labour force. In the 2017 Seychelles strategy published in 2007, the

Table 9.2. Key policy and strategies. (From STB, 2012; STA, 2015.)

Source	Main components
Vision 21 (2001–2010)	Introduced the concept of ecotourism Encouraged tourism businesses to adopt environmentally friendly and energy efficient practices (STB, 2012)
Tourism Master Plan	Proposes the vision for tourism in Seychelles as a sustainable destination Reviews the main trends in the Seychelles tourism industry Discuss the importance of mainstreaming tourism and environment Recognize the importance of policy frameworks for tourism Shows the main sectors/business of the tourism industry Recognize limitation in telecommunication but does not mention social media Recognized the very limited supply of human resource Proposes future projects to promote cultural aspects of Seychelles such as heritage centre and museum
Seychelles Tourism Academy Strategic Plan (2015–2019)	Proposes various strategies in terms of human resources, financing and quality assurance in order to deliver more graduate courses in tourism and hospitality to the industry Has a vision 'To be recognized as the leading hospitality and tourism training institute in the region, reputed for its programmes on offer and the quality of its graduates' Plan of the new academy to be built in different phases to cater for a higher number of trainees

Table 9.3. Comparison of economic performance of Seychelles, Mauritius and Maldives. (From WTTC, 2015a, b, c amended by the authors.)

	Seychelles	Mauritius	Maldives
GDP: Direct contribution	21.3% of GDP	11.3% of GDP	41.5% of GDP
GDP: Total contribution	56.9% of GDP	25.5% of GDP	78.1% of GDP
Employment: Direct contribution	22.9% of total employment	10.9% of total employment	32.2% of total employment
Employment: Total contribution	57.3% of total employment	24.2% of total employment	62.0% of total employment
Visitor exports	32.3% of total exports	24.5% of total exports	76.2% of total exports
Investment	29.5% of total investment	6.0% of total investment	18.9% of total investment

government had identified tourism along with fishing as the engine for growth of the economy and the target set was doubling the income per capita from $8,722 in 10 years. This target while sounding ambitious is achievable as the country's GDP per capita in 2014 already amounted to $15,359.20. The economic success of the islands are to a large extent attributable to the rapidly increasing number of visitors to the islands from 122,038 in 2003 to 232,667 in 2014.

TTCI: a snapshot of Seychelles

In 2015, the archipelago was ranked fifty-fourth out of 141 economies in the WEF Travel and Tourism Competitiveness Index (TTCI) ranking. In regional terms, it is second to South Africa in the sub-Saharan region and first amongst the Indian Ocean islands (Table 9.4). The two areas where Seychelles perform extremely well are the prioritizing of the tourism sector by the government and the nation's commitment to environmental policies. As mentioned in the section above, from the 2001, the government laid out its plan for a more sustainable development in the Vision 21 document (Table 9.2). Environmental management in Seychelles has been commended by academics such as Gossling and Horstmeier (2003), who acknowledge the commitment of the Seychelles government in that regard. In addition to governmental bodies and authorities, non-governmental organizations (NGOs) such as Nature Seychelles and Seychelles Island Foundation (SIF) are important custodians of the environment in Seychelles. Tourism revenue is regarded as a significant source for continued conservation (Steven *et al.*, 2013). The heightened political attention to the environment according to Edgell and Swanson (2013) is the economic growth that is linked to tourism. Data from NBS suggest that snorkelling and hiking and nature tours are the preferred activities of visitors. This implies that the natural environment of Seychelles is fundamental to its tourism product.

Table 9.4. Competitive sub-indices. (From WEF, 2015, p. 4.)

Sub-index	Indicators	Seychelles
Enabling environment	Business environment	48
	Safety and security	86
	Health and hygiene	60
	Human resources and labour market	59
	ICT readiness	66
Travel and tourism policy and enabling conditions	Prioritizing of travel and tourism	8
	International openness	87
	Price competitiveness	123
	Environmental sustainability	6
Infrastructure	Air transport infrastructure	24
	Ground and port infrastructure	29
	Tourist service infrastructure	15
Natural and cultural resources	Natural resources	85
	Cultural resources and business travel	138
Overall ranking out of 141 economies		54

On the other hand, two areas where the country fared quite poorly are price competitiveness, cultural resources and business travel. As the country is not currently seeking to diversify into business tourism, the poor ranking is not of immediate concern to the authorities but these can be seen as opportunities for diversification for future development. However, the Tourism Master Plan clearly states that one of its objectives is to develop cultural tourism to the islands. The colonial heritage of the islands, their role in the history of settlement in the Indian Ocean, multiculturalism of the population and the high quality of traditional food can be the basis for developing this type of tourism on the islands. However, the emphasis on promotional videos is still sun, sand and sea. Places of historical importance and the studios of local artists are listed as attractions on the website of STB. The STB is also promoting the Kreole Institute (Kreole is the most spoken language in Seychelles). This centre has a cultural exhibition, language and literature workshop and celebrates the French heritage of the islands by organizing recitals of traditional French songs.

The immediate concern for the industry, however, is the low ranking obtained for the price competitiveness. As a remote destination with limited air access, the cost of travel to Seychelles is already high. Cost of living is high due to a high content of import of the local consumption and this means that the tourism product is costly too. The prices may also indicate the archipelago's successful strategy of exploiting a certain degree of market power that it may enjoy in the region by providing unique tourism products associated with inelastic demand, for example the endemic fauna and flora. This allows the market to grow while keeping prices high. The high and sticky prices may be indicating the oligopolistic nature of tourism providers as suggested by Lopez-Guzman *et al.* (2013), who states that tourism ownership in African countries is predominantly held by a small number of international companies. The Seychelles government, however, attempts to alleviate this by allowing the provision of certain services such as taxi operations and tourist guides solely by local residents only, although luxury establishments remain predominantly foreign owned (STB, 2012). However, growing competition in the region from new destinations such as Sri Lanka and existing destinations such as Maldives and Mauritius may mean that in the future, the destination will need to be more careful in its pricing, which will be a hard balance to achieve given the high content and cost of import in the provision of the tourism product.

Tourist arrivals

Arrivals in Seychelles have risen from a few thousands in the 1970s to 240,000 in 2014 (Fig. 9.1). The spectacular increase reflects the commitment of the country to this industry although the growth has not always been steady.

In spite of the specular growth from 1970 to 2014, the Seychelles islands have experienced periods of decline in international arrival figures attributable to global economic conditions. The industry, however, recovered and has continued to grow. Table 9.5 lists the events that account for the trends in arrivals in Seychelles from 1970 to 2014. According to Vannier and Uranie (2015) the single most important factor has been improvement in air access.

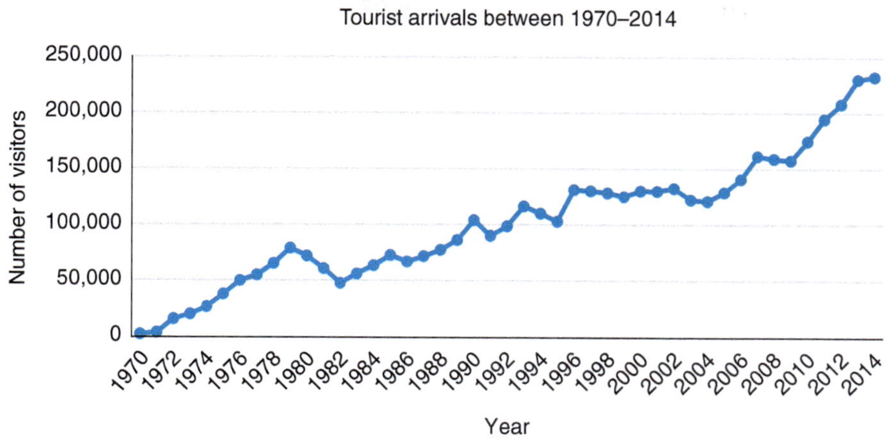

Fig. 9.1. International arrivals to Seychelles from 1970 to 2014. (From NBS, 2015b.)

Table 9.5. Trends in tourist arrivals. (From Gabbay and Ghosh, 2003; STB, 2012.)

Year	Trend	Explanation
1971–1979	Drastic rise	The airport opened in 1971 hence allowing for greater flow of tourists This rise occurred despite the military coup in 1977
1979–1983	Decline	1979: beginning of energy crisis 1979–1985: cold war From 1980: global recession
From 1987	Rise in tourist numbers	1989: regular flight between Malaysia and Seychelles resumed
1990	Sharp peak	Acquisition of a new Boeing 767 to make it the third in Air Seychelles' fleet (The sharp rise and fall could be due to temporarily discounted fares)
1992–1993	A sudden rise then peak in 1993	Introduction of the multi-party state improved the perception of the destinations as stable Increased construction of guesthouses 1992: increase in number of South African travellers following an 11 years' ban lift (this year links to end of apartheid in South Africa)
2007–2009	Decrease	Global economic crisis
2009–2013	Sharp increase	Increased number of flights by Etihad Airways and Emirates Despite the suspension of direct flights to Europe by Air Seychelles, the tourism numbers continue to rise. In 2012 Air Seychelles entered into codeshare agreement with Etihad hence increasing seat capacity to the destination

There may also be other reasons for the continued growth in recent years, such as word of mouth. Seychelles continue to deliver high-quality products. For instance, according to the National Bureau of Statistics (2014b), 90% of visitors would recommend Seychelles. Furthermore, in the 2015 regional World Travel Awards, the archipelago obtained the best: airline, cruise ship port in addition to several of the prizes in the accommodation category in the Indian Ocean region. Arguably, the increasing visitor numbers could warrant careful consideration and planning in future. Figure 9.2 compares arrivals to the population.

By 2011 the number of arrivals to Seychelles was already twice as high as the population of the island, which has not undergone significant changes in the last decade. The outcome of the situation is that in the future, labour shortage may become a limiting factor in the development of the tourism industry, which is highly labour intensive and service driven. Recent research by Philpot *et al.* (2015) brings to light vulnerabilities of Seychelles but from the perspective of the local people. The research highlighted the lack of skilled as well as qualified personnel as one of the key contributors to social vulnerability of the island. The shortage of labour is equally highlighted in the Tourism Master Plan. The government proposes to build an academy that will offer training and graduate level courses in tourism and hospitality. This is will improve the quality of the labour force and educate and empower the local residents, enabling them to join the industry. However, in the long run given the small population of the islands, the problem of shortage of labour may persist. Immigration and hiring short-term labour to fill in skill gaps, which is already being implemented on the islands, may offer some relief. However, employing foreign workers has the potential of exacerbating the problem of leakages in the form of remittances to the home country in a country where economic leakage is already quite high.

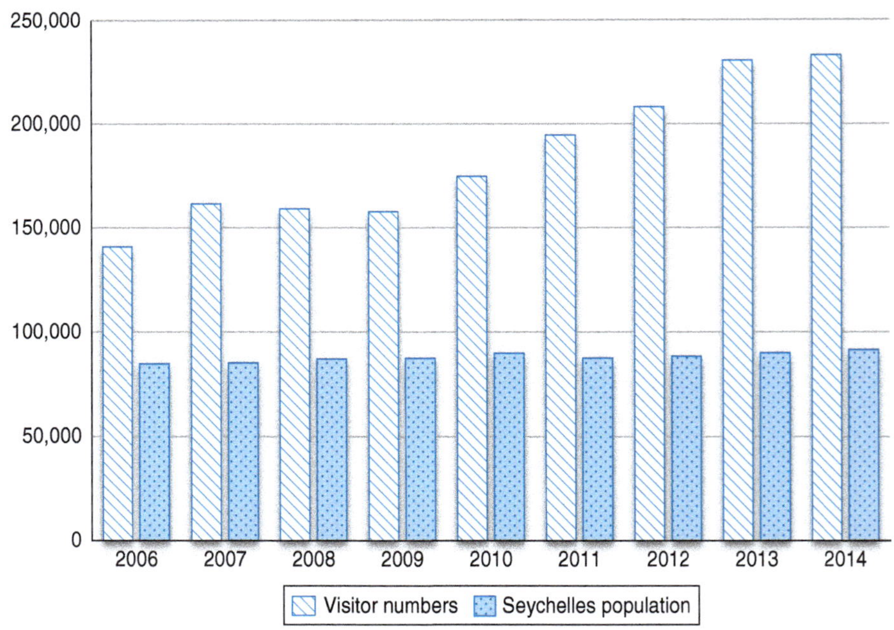

Fig. 9.2. Visitor numbers compared to local population. (From NBS.)

Market mix and air access

The remoteness of the island is recognized by Gabbay and Ghosh (2003) as a unique selling point in terms of marketing. However, they argue that as a consequence of its remoteness and exclusion from other countries, transportation as well as communication with the rest of the world remains problematic. Like most SIDS, Seychelles relies on long-haul travellers. Until 2013, the three main markets were France, Italy and Germany representing almost 50% of total arrivals (Fig. 9.3). However, the proportion of arrivals from Italy has declined steadily while that of Germany has been growing slightly on average by 0.5%. From 2011 onwards however, the market mix has changed. The proportion of arrivals from France declined sharply. According to Hampton and Jeyacheya (2013) this may have resulted from reduced access to the archipelago during the time period.

For example, direct flights from France were cancelled in 2012 and Vannier and Uranie (2015) state that French travellers are less likely to book flights to destinations that are not serviced by direct flights, which may have resulted in the fall in the number of arrivals from this market. New routes, however, were created linking Seychelles to the Middle East and China. As a result China and UAE have gained in market shares and become prominent sources of arrivals for Seychelles, generating 8% of total arrivals. However, by 2015, a declining European market and its consequences on the Seychelles tourism industry, led to direct flights being resumed to Paris. Currently, Air Seychelles offers three direct flights to Paris per week. This has given a boost to the French market, which is again the main source of arrivals to the destination. This episode highlights the importance of air access to the destination.

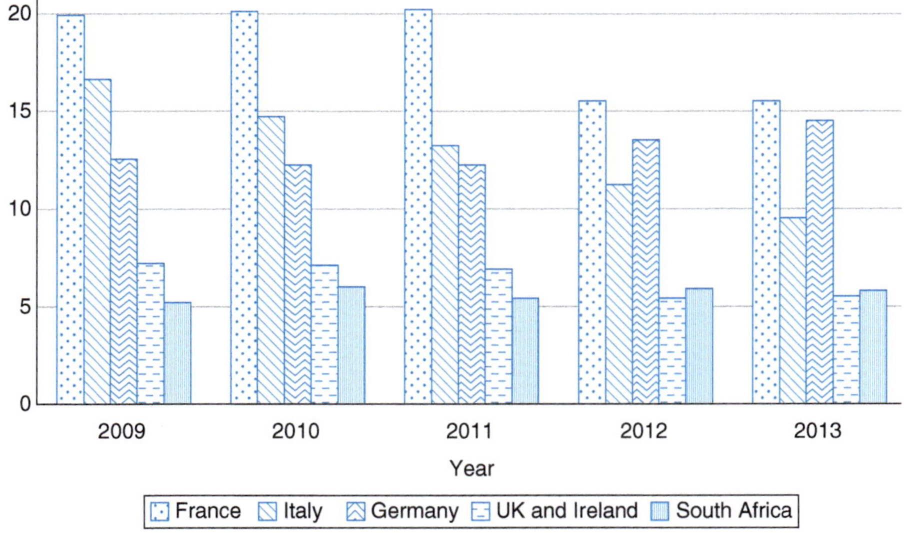

Fig. 9.3. Market and air access to Seychelles. (From NBS data compiled by the authors.)

9.3 Tourism in Seychelles – The Consumer's Perspective

Ramis and Prideaux (2013) argue that visitors' perceptions are influenced by whether they receive positive or negative information communicated to them through marketing or the media. Similarly, it is argued that travel planning is increasingly being influenced by strangers on the Internet (Kladou and Mavragani, 2015). In order to understand the consumers' perspective of tourism in Seychelles and identify their most important concern, qualitative data from TripAdvisor (TA) were analysed and these were substantiated with the qualitative answers recorded in the NBS exit survey. NBS is responsible for analysing and presenting this information. The data forms part of the Visitor Safety and Security Survey bulletin. Information from the year 2014 is utilized to inform this research.

TA is a platform that allows travellers to advise others on trips as well as plan visits to various attractions. It is the largest travel site and contains more than 225 million discussions. This is a method for qualitative data on customers' opinions of destinations and services using an online platform (Rageh *et al.*, 2013). TA is considered one such site, prominent for reviews and forum interactions (Buhalis and Law, 2008), hence contributing to the concept of electronic Word of Mouse (eWOM). In fact it is argued that the comments placed there can possibly go a long way in influencing travel decisions. In view of all of the above points, it was deemed a favourable choice as a means of information for the purpose of this study. In this case community interaction was observed and collected for the whole of 2014. The findings from the data collected from TA were compared to qualitative data collected through the Tourism Exit Survey of the NBS. The self-administered questionnaire is administered to tourists quarterly.

The initial findings suggest that perception of the tourists is indeed multidimensional as claimed by San Martin and Rodriguez del Bosque (2008). The perception is considered to be predominantly cognitive (related to the features of the destination) rather than affective (related to the feeling). The environment is one key functional cognitive attribute seen. This can also be associated to the physiography and climate core resources and attractors Crouch and Ritchie (1999) discussed earlier. The findings are supported by those from Kladou and Mavragani (2015) who assert that it is the cognitive component that attracts the most comments on TA.

The tourists' attraction

As discussed by Cameron and Gatewood (2008), the findings show that the main strength of Seychelles remains its key tourism product, pristine beaches and the fauna and flora. It is featured in both the TA and the NBS data analysed. The visitors are highly complementary about the island natural tourism capitals. The experience of the visitors is enhanced through the fact that there are a large number of beaches to choose from. Furthermore, because these are located on different islands, tourists are spread out minimizing risk of concentration and crowding. The tourists' positive comments on the environment show the success the government's commitment to conserving the natural environment of Seychelles.

In general the NBS survey shows that tourists find the locals to be very friendly and hospitable. However, this is in contrast to Philpot *et al.*'s (2015) study of vulnerability and resilience, which discussed locals' unhappiness at being denied access to certain beaches in order to uphold tourist privacy. The discrepancy could mean that, although there is a certain degree of resentment, it has not reached the level whereby the locals exhibit it directly to the tourists. Food served was generally perceived positively in Seychelles and the Victoria market was deemed a must-see for visitors. The unique local dishes reflect the ethnic diversity of the country. The local population originate from Africa, Asia and Europe. The blend of culture has inspired the local cuisines that are very much appreciated by the visitors. The positive attitude toward the unique cuisine of the island may indicate the potential for developing culinary tourism as a niche market.

Costs

Regarding weaknesses, the main problem remains the cost of the holiday. '*Seychelles is an expensive destination*' is echoed throughout the discussion forums on TA. It is not surprising that potential visitors are choosing to shop around and seeking more information on options available to them for example, one visitor asked the forum, '*The price is 490 Euro which seems a little too high, does anyone have idea about this?*' Another adds that '*I think considering the costs that some of the tour companies are quoting, it's better to do this on my own*' or '*I thought there would be alternative options offering the same service at a lower price?*' These perceptions indicate that the message of the 'Affordable Seychelles' campaign discussed in the Master Plan has not yet been successful in changing the perception of Seychelles as a top luxury market. Nevertheless, during collection and analysis of the data it was noted that a couple of tourists agreed that Seychelles could be experienced on a budget.

Service quality

However, high prices need not be a problem if they are perceived as value for money. Unfortunately this is not always the case. The varying level of satisfaction of consumers from smaller businesses may become a serious cause for concern and have implications for the destinations' image. It can indicate a lack of strategy on ensuring product quality or poor implementation of policies. Several consumers claim that the price charged for example '*Taxis are very expensive in Seychelles so expect to pay more than it is worth*'. In general, the tourists have experienced different level of quality of services. While one couple offer excellent feedback on their guide '*...and we were very satisfied. He is not just driving you from A to B, but he guides you, shows you the most interesting things on the island and gives you a lot of information and tips. He is very friendly, reliable and flexible, and I would recommend him to anyone*', others are complaining about the quality of service and products offered. For example about a pizzeria, customers complain that '*service quality is mediocre*'. As of 2016, taxi meters have been re-implemented in order to ensure a more standardized tariff. This shows the commitment of the authorities to address some of the limitations related to tourism.

Transportation in Seychelles

Linked to the concerns raised about prices and service quality, are issues related to transportation. Not only is Seychelles costly to access from its main market as discussed before, transport links within the archipelago are perceived as being expensive as well. Prospective tourists are often concerned about the lack of diversity and choice in transportation link to other islands. It is mentioned that, *'From Mahe to Praslin, 100 Euro sounds a lot! So how much would the flight cost?'* Note this trip takes 1 hour each way. The lack of choice between ferry and domestic flights means that visitors have to pay the fare, making it an expensive addition to the budget especially if their accommodation is not located on Mahé. Regarding land travel in the island, the following were mentioned: *'Buses are very safe but they are not very reliable'* and *'but you cannot count on buses for dinner in Seychelles. Last bus leaves Victoria at 7.30 pm'*. Given the size of the local population, offering more frequent services may not be profitable due to lack of density. Extending bus hours may only be possible through subsidies from the government. In the Tourism Master Plan, however, the government's aim is to encourage tourists to use taxis and rental cars.

Communication technology and information

As mentioned before, the World Economic Forum ranked Seychelles sixty-sixth for ICT readiness (WEF, 2015) suggesting room for improvement. Lack of up to date communication technology was picked up in the Tourism Master Plan, following feedback from consumers from the NBS survey, on the need for better connectivity. In the TA forums, a range of questions were asked on the availability of Wi-Fi, Internet and the cheapest way to make international calls. Criticisms were made on the limited possibilities for making purchases online. Several of the smaller businesses did not have websites or provide email addresses. These were true mostly for the providers of accommodation. When websites were available, they lacked 'call to action' links and information provided were not always adequate. This can mean that tourists are unable to book the cheaper accommodation provided from their home country. This may discourage potential visitors on a budget from visiting the island and represent a loss of revenue for the smaller businesses. The TA echoes many of these concerns whereby prospective visitors feel information on the destination is not readily available for example *'thank you so much for your reply ... it is so hard to find any information on the Seychelles regarding self-catering'*.

In fact during the data collection period, the researchers tried to gather information from the 'Seychelles secrets' website, which regroups small Seychelles establishments, but it was inaccessible for a fairly long period of time. This webpage is one avenue to provide the tourist with information. NBS data included complains about lack of information on trails, weather forecasts and information about various other businesses. One customer states that, *'we're planning a trip to the Seychelles for my husband's birthday but there are limited agencies in the US who have expertise in the Seychelles'*. Lack of access to information occurred in the planning phase of the trip and during the trip in Seychelles. Provision of good information seems to be an element that requires several improvements. The Seychelles Tourism Board website (Seychelles Tourism Board, 2014) for instance was hardly ever mentioned as a recommended reference amongst TA users, despite the wealth of information that it already provides.

Health and safety

As stated by Hall *et al.* (2003), feeling safe is an important element in a visitor's satisfaction. While in general the consumers seem fairly satisfied with the level of safety, nevertheless a few had issues that they discussed. It is not surprising to find a potential tourist asking, *'Can anyone recommend anywhere and is it safe to walk around in the evening?'* Prospective tourists seem to be concerned about public transport conditions and risk of thefts and other forms of attack. The answers received were reassuring implying that Seychelles was regarded as a fairly safe destination in this respect. Other concerns were related mainly to the sea and the beaches. Seaweed seems to have been a problem not just for aesthetic reason but tourists who had concerns regarding potential health hazards. Others were worried about sandflies and other insects and access to medical facilities in case of accidents. For example, one person asked, *'Does anybody know whether there are sandflies on the beach ... ? My son has an allergy for insect bites, so it could ruin our whole holiday'.* Issues regarding reading health and safety at sea were raised too as one person commented, *'Most of the tourists do not stay inside the net because on the right side of the bay there are lots of fish and there are chances also to spot turtles'.* These concerns were perhaps amplified by the memories of a shark attack that took place in 2011 and which was highly publicized in the international media. For example one enquirer on TA wished to avoid La Reserve beach thinking this was the location of the attack. More knowledgeable past tourists were instrumental in confirming that the attack was instead at Anse Lazio. Others were also quick to reassure potential visitors. One person stated that he had swum there and that nets were in place, and there have not been any attacks since. This point also links to the one on how TA can help to influence the decisions of other travellers. Their experience can in a way be more convincing than the words of the authorities.

Competition

Finally, it is interesting to note that in spite of significant investment in promotion campaigns, Seychelles does not seem to have been able to distinguish itself as a unique destination in the eyes of some prospective tourists. It is in close competition with other islands in the Indian Ocean who offer similar products for example, Mauritius. Several potential customers seem to be undecided between Seychelles and Mauritius and they have the following to say: *'... too confused between destination Seychelles and Mauritius', 'My husband and I are thinking about visiting Mauritius (Beachcomber Paradis Hotel) or Seychelles (Constance Ephelia) later this month with our daughter.'* and *'I would prefer Mauritius whereas my husband would prefer Seychelles. Both look lovely but get the impression that Seychelles is much quieter'.* This called for more emphasis on the uniqueness of Seychelles as a destination. The fact that Seychelles is outside of the cyclonic zone of the Indian Ocean seems to have a positive influence on the choice of destinations. The average stay of visitors in the Seychelles is around

10 nights (NBS, 2015c). Bad weather for more than a couple of days will have a very significant negative effect on the trip as a whole.

9.4 Conclusion and Recommendations

The aim of this chapter was to provide additional information in the form of consumers' feedback that may inform policy making. The netnographic approach was applied to collect data from discussion forums such as TripAdvisor to better understand consumers' opinions and queries on the tourism product of Seychelles. These were compared to the qualitative data collected by the National Bureau of Statistics of Seychelles. It is evident from the data collected that Seychelles is a highly prized destination. Consumer feedback on the quality of the environment gives credit to the government's commitment to placing a high importance on the preservation of the natural beauty of the archipelago. The additional measures declared in 2015, which include measures to reduce pollution on the island is likely to be well received by the visitors.

There are nevertheless a few areas that could be further improved. One such area that requires immediate attention is access to information. The salient point of this study is that tourism is more and more reliant on social media as a source of information, and platforms such as TripAdvisor are fast becoming the first point of call. While the STB does have Facebook and Twitter accounts, the engagement is quite minimal. The level of engagement with social media, however, needs to increase. One suggestion is that the STB participate actively in the discussion forum on TripAdvisor. It can be used to direct consumers to the Facebook and Twitter accounts where live and up-to-date information can be provided and to answer queries. Regular analysis of data from the TA site can help the STB identify areas of problems that tourists may be facing. These platforms may be integrated in the marketing strategies of the STB. Real-time answers to queries and negative comments can help influence consumers' satisfaction in a positive manner and prevent the spread of incorrect information on the destination. For example, budget-minded consumers may be pointed to the cheaper accommodation options, to further promote the destination as an affordable one.

This can be further facilitated through investment in the IT services and training on the islands. Finally, in a bid to remain competitive, the tourism industry should consider professional and uniform service across all the islands. Quality control of smaller establishments where the service provided has been perceived as of adequate standard is required to ensure that tourists perceive the service they received as value for money. The tourism academy proposed in the Master Plan should alleviate some of the problems. A trained labour force can be expected to better understand and cater for the needs of the tourist. It is suggested that the academy also provide training in ICT and related areas to small businesses to help access social media and better market their products on websites.

The findings of this piece of research are relevant for other small island states that face similar challenges to Seychelles, such as Mauritius and Maldives. Small islands have limited resources both in term of finance and human capital. Remoteness, which is a strength for their tourism industries, adds to the budget of the consumers. Several of the issues discussed by the tourists in Seychelles such as lack of a standard quality of services, limited ICT-related infrastructure and access to real-time information are relevant for Mauritius. This makes a very strong case for the monitoring of widely accessed online discussion forums. These forums provide more detailed and specific information than exit surveys and have a wider reach that includes both visitors and potential visitors. The information provided can be used to identify areas of concern and improve service quality. Finally, as Tosun and Jenkins argue (1998), planning approaches may sometimes take developed countries approaches to impose on developing countries. The country needs to continue to adopt more endogenous strategies that are more tailored to the specificities of Seychelles as a SIDS, but also the specific attributes the differentiate it from other small islands.

References

Buhalis, D. and Law, R. (2008) Progress in information technology and tourism management: 20 years on and 10 years after the Internet–The state of eTourism research. *Tourism Management* 29(4), 609–623.

Cameron, M.C. and Gatewood, B.J. (2008) Beyond sun, sand and sea: the emergent tourism programme in the Turks and Caicos Islands. *Journal of Heritage Tourism* 3(1), 2008.

Croes, R. (2006) A paradigm shift to a new strategy for small island economies: embracing demand side economics for value enhancement and long term economic stability. *Tourism Management* 27(3), 453–465.

Crouch, I.G. and Ritchie, B.R.J (1999) Tourism, competitiveness, and society prosperity. *Journal of Business Research* 44(3), 137–152.

Edgell, L.D. and Swanson, R.J. (2013) *Tourism policy and planning: yesterday, today and tomorrow*, 2nd edn. Routledge, Oxford, UK.

Filieri, R. and McLeay, F. (2014) E-WOM and accommodation: an analysis of the factors that influence travelers' adoption of information from online reviews. *Journal of Travel Research* 53(1), 44–57.

Gabbay, R. and Ghosh, R. (1997) *Tourism in Seychelles.* University of Western Australia, Nedlands.

Gabbay, R. and Ghosh, R. (2003) Tourism in the Seychelles. In: Ghosh, R., Siddique, M. and Gabbay, R. (eds) *Tourism and Economic Development: Case Studies from the Indian Ocean Region.* Ashgate Publishing, Farnham, UK, pp. 104–127.

Gossling, S. and Horstmeier.O. (2003) 'High-value conservation tourism': integrated tourism development in the Seychelles? In: Gossling, S. (ed.) *Tourism and Development in Tropical Islands: Political Ecology Perspectives.* Edward Elgar Publishing, Cheltenham, UK, pp. 203–221.

Hampton, P.M. and Jeyacheya, J. (2013) *Tourism and Inclusive Growth in Small Island Developing States.* Commonwealth Secretariat, London.

Hampton, P.M. and Jeyacheya, J. (2015) Power, ownership and tourism in small islands: evidence from Indonesia. *World Development* 70, 481–495.

Kladou, S. and Mavragani, E. (2015) Assessing destination image: an online marketing approach and the case of TripAdvisor. *Journal of Destination Marketing & Management*, in press.

Kozinets, V.R. (2010) *Netnography: Doing Ethnographic Research Online.* SAGE, London.

Laporte, P. (2013) *Budget Speech 2014.* Ministry of Finance Trade and Investment, Victoria, Seychelles.

Lopez-Guzman, T., Borges, O., Hernadez-Merino, M. and Maria Cerezo, J., (2013) Tourism in Cape Verde: an analysis from the perspective of demand. *Tourism Economics* 19(3), 675–688.

Mintel (2014) *All Inclusive Holidays – June 2014.* Mintel Group, London.

National Bureau of Statistics (2014a) *Seychelles in Figures: Various Issues.* Victoria, Seychelles.

National Bureau of Statistics (2014b) *Visitor Expenditure Survey, Quarter 3–2014.* Victoria, Seychelles.

National Bureau of Statistics (2015a) *Population and Vital Statistics: December 2014.* Victoria, Seychelles.

National Bureau of Statistics (2015b) *Visitor Arrival Statistics: Various Issues.* Victoria, Seychelles.

National Bureau of Statistics (2015c) *National Bureau of Statistics Statistical Bulletin: Visitor Safety and Security Survey: Various Issues*. Victoria, Seychelles.

National Bureau of Statistics (2015d) *Migration & Tourism Statistics*. Victoria, Seychelles.

Philpot, D., Gray, S.T. and Stead, M.S. (2015) Seychelles, a vulnerable or resilient SIDS? A local perspective. *Island Studies Journal* 10(1), 31–48.

Rageh, A., Melewar, C.T. and Woodside, A. (2013) Using netnography research method to reveal the underlying dimensions of the customer/tourist experience. *Qualitative Market Research: An International Journal* 16(2), 126–149.

Ramis, M. and Prideaux, B. (2013) The importance of visitor perceptions in estimating how climate change will affect future tourist flows to the Great Barrier Reef. In: Reddy, V.M. and Wilkes, K. (eds) *Tourism, Climate Change and Sustainability*. Routledge, Oxford, UK, pp. 173–188.

Rudestam, E.K. and Newton, R.R. (2015) *Surviving your Dissertation: a Comprehensive Guide to Content and Process*. 4th edn. SAGE, Los Angeles, California.

San Martin, H. and Rodriguez del Bosque, A.I. (2008) Exploring the cognitive-affective nature of destination image and the role of psychological factors in its formation. *Tourism Management* 29, 263–277.

Seychelles Government (1997) *Seychelles Strategy 2017*. Seychelles Government, Victoria, Seychelles.

Seychelles Tourism Academy (2015) *STA Strategic Plan: 2015–2019*. Seychelles Tourism Academy, Victoria, Seychelles.

Seychelles Tourism Board (2012) *Seychelles 2012-2020 Tourism Master Plan: for a Sustainable and Responsible Tourism*. STB, Victoria, Seychelles.

Seychelles Tourism Board (2014) *The Seychelles Islands Another World: Promotional Brochure*. STB, Victoria, Seychelles.

Steven, R., Castley, G.J. and Buckley, R. (2013) Tourism revenue as a conservation tool for threatened birds in protected areas. *PLOS one*. Available at: http://journals.plos.org/plosone/article?id=10.1371/journal.pone.0062598 (accessed 15 August 2015).

Tosun, C. and Jenkins, L.C. (1998) The evolution of tourism planning in third-world countries: a critique. *Progress in tourism and hospitality research* 4, 101–114.

TripAdvisor (2015). Destination experts forum. Available at: https://www.tripadvisorsupport.com/hc/en-gb/articles/200613327-Who-are-the-destination-experts-on-the-forums- (accessed 24 August 2015).

Vannier, R. and Uranie, S. (2015) France is back as the Seychelles main tourism market. *Seychelles News Agency*. Available at: http://www.seychellesnewsagency.com/articles/2397/France+is+back+as+the+Seychelles+main+tourism+market (accessed 15 August 2015).

World Bank (2017) *World Development Indicators*. World Bank, Washington, DC. Available at: https://data.worldbank.org/country/seychelles (accessed 9 October 2017).

World Economic Forum (2017) *The Travel & Tourism Competiveness Report 2015: Growth through Shocks*. WEF, Geneva, Switzerland.

World Travel & Tourism Council (2015a) *Travel & Tourism: Economic Impact 2015: Seychelles*. WTTC, London.

World Travel & Tourism Council (2015b) *Travel & Tourism: Economic Impact 2015: Mauritius*. WTTC, London.

World Travel & Tourism Council (2015c) *Travel & Tourism: Economic Impact 2015: Maldives*. WTTC, London.

10 Destination Management: A Network Perspective

Michelle McLeod[1]* and Noel Scott[2]

[1]*The University of the West Indies, Mona, Jamaica;* [2]*Griffith University, Southport, Australia*

10.1 Introduction

A tourist destination is the location that tourists physically visit and, depending on the travellers' dispersion, may be a country, island, state, area or particular resort (Ritchie and Crouch, 2003). Destinations that encompass many attractions, independent businesses and that receive significant tourism flows require management and coordination for activities such as collaborative marketing. The entity providing governance of a destination is termed a Destination Management Organization (DMO) and the study of such entities is a mainstream topic in the tourism literature (Laws, 1995; World Tourism Organization, 2007; Scott, Cooper and Baggio, 2008; Pearce, 2015).

A DMO can help to build successful tourism destinations through increasing promotion, developing new tourism experiences, encouraging quality improvements and especially through providing a basis for collaboration. Such collaboration is important since a satisfactory destination experience is co-produced through the inputs of accommodation, attraction and transport operators interacting to deliver tourism products and services. An understanding of the extent and type of stakeholders' interaction within a destination can assist with improving destination management practices. Study of stakeholder networks of interaction also identifies the central agencies involved in the coordination and development of the destination. In this research study the core tourism activities within a small island destination, and the related stakeholder relationships, are explored.

Destination networks in the context of the tourism sector are formed from the interactions of tourism stakeholders or actors as they deliver tourism products and services. Networks and their evolution as a destination system have been explored by several authors (Scott, Cooper and Baggio, 2008; Pavlovich, 2014; Pforr, Pechlaner, Volgger and Thompson, 2014; Baggio and Sainaghi, 2016). Network analysis has

*E-mail: michelle.mcleod@uwimona.edu.jm

also been used to study tourist movements based on transport dynamics (Lew and McKercher, 2006) and mobility dynamics (D'Agata, Gozzo and Tomaselli, 2013).

The purpose of this chapter is to explore, using a case study approach and network analysis, the coordination of tourism activities within a destination and thereby highlight the importance of destination management collaboration practices for tourism destination success. This exploratory case study is based on a small island tourist destination that is in decline in terms of visitor numbers and expenditure. An island bounded by sea is an ideal context to study tourist activity, particularly if the tourism sector is a mainstay of economic activity. This case contributes to research concerning the destination life cycle and destination sustainability (Butler, 2006) and in particular to the destination decline stage (Manente and Pechlaner, 2006). Exploration of the case study of the destination using network analysis may provide recommendations that can turn decline into rejuvenation.

10.2 The Concept of a Tourism Destination Network

Although the term destination has been used to describe a concentration of tourist activity, there are several avenues that can be taken to conceptualize the nature of a tourism destination. Pearce (2014) notes an integrative framework to conceptualize a destination and points to industrial districts, clusters, networks, systems and social constructs as conceptual frames. Pearce (2014) suggests that the operation of tourism and hospitality industries as destinations can be viewed as supply and demand systems as with the operation of a network. Other tourism researchers have supported the conceptualization of destinations as networks (Hristov and Zehrer, 2015). A destination network is the composition of suppliers of tourism products and services that are coordinated to deliver the tourism experience.

This relational view of a tourism destination can add value to destination research. Laesser and Beritelli (2013) view destinations as geographic entities with a network of suppliers providing for tourist demands. A geographical approach assists with understanding the destination as a spatial construct that is based upon attraction products. Attractions are important to building a destination's comparative advantage (Ritchie and Crouch, 2003) and as attractions pull tourists to the destination, a destination network forms to produce tourism products and services to satisfy tourists. A simple geographical spatial view alone is limited since by its very nature, a tourism experience is built from a network process as several suppliers add value to deliver the total tourism activity. Understanding the nature of the network process within tourism destinations can facilitate improved management of the destination. In fact, destination management can be viewed as network management (see Laesser and Beritelli, 2013) and therefore, the concept of destination management within tourism should involve consideration of the characteristics of a tourism destination network.

Tourism destination network management, coordination and handling crises

Primarily destination network management involves coordination. This coordination occurs through both formal and informal interventions and influences of the DMO and other stakeholders in the tourism sector. Volgger and Pechlaner (2014) point to the critical importance of the DMO in network management. DMOs are central conduits to 'promote self-responsibility, self-organization and self-regulation of the destination network' (p. 64).

There is one critical question. Can a destination's tourism activities be sustained and managed without the intervention of a DMO? Volgger and Pechlaner (2014) understand the important role of networking for destination management and supports destination networks as governance structures to develop joint strategies and promote collective action. Similarly, Hristov and Zehrer (2015) support the important role of the DMO as a networked platform for nurturing common thinking through knowledge sharing and collective action. Volgger and Pechlaner (2014) argue that the networking capability of the DMO is significant for its acceptance among stakeholders. The range of tourism activity within the destination may be beyond the purview of a DMO. The destination suppliers have collective interests in the provision of tourism products and services and these interests are supported through the inter-organizational links between stakeholders. The ongoing network of tourism activity has to be coordinated to build a successful tourism destination and arguably while the DMO has a role, there is need to understand how the destination network acts as a coordinating mechanism.

Destination coordination becomes of especial importance when there is a crisis within the destination. Crises are events that challenge the normal process of tourism activities within the destination and can have major impacts on its economic, environmental and social aspects. As tourist demand for destination products and services change in time (Baggio and Sainaghi, 2016) and space (D'Agara, Gozzo and Tomaselli, 2013), capacity management during times of crises becomes important. Transformation of a destination network during a crisis event (Becken, Scott and Ritchie, 2015) and the emergency management processes that occur during destination crises (Morakabati, Page and Fletcher, 2016) are important considerations, but more so, the capability of a DMO to serve as a crisis coordinator may be explored by study of a destination's network characteristics.

Tourism destination development and decline

Ritchie and Crouch's (2003) work provided a seminal exploration of how a tourism destination achieves competitiveness and sustainability. Their conceptual model of destination competitiveness includes factors such as: system definition, philosophy/values, vision, positioning/branding, development, competitive/collaborative analysis, monitoring and evaluation and audit (Ritchie and Crouch, 2003, p. 63). Destination development activities are primarily the responsibility of a DMO to ensure that the required destination performance is achieved; however, there are other conceptual approaches for study of tourism destination development.

Destination networks act as facilitators for tourism destination development. Pavlovich (2014) explored destination network evolution noting that the process of tourism destination development can be considered as increasing connectivity within the destination network. Numerous connections among tourism operators formed through product synergies, decision-making activities, employees' activities and social ties and knowledge transfer support the transformation of the destination network (Pavlovich, 2014). Komppula (2014) challenges the role of the DMO in destination development and focuses on the collaboration that occurs between the tourism businesses of rural destinations where network relationships are sparse and decentralized. Tourism business activity supports destination development and destination development depends on the direct and indirect ties of actors that influence a destination network (Strobl and Peters, 2013). Nonetheless, the activities of a DMO that contribute to tourism development can be explored using a network perspective to analyse whether the tourism development capability of the DMO is constrained by certain network characteristics.

Within the tourism literature, Manente and Pechlaner (2006) and Diedrich and Garcia-Buades (2009) have studied tourist destination decline. These researchers explored Butler's (1980) Tourism Area Life Cycle (TALC) to reveal the internal characteristics that would lead to the decline stage in the destination life cycle. Diedrich and Garcia-Buades (2009) considered the local perceptions of tourism as a key factor, while Manente and Pechlaner (2006, p. 248) provided a review of indicators of destination decline, such as residents' ratio, excursionist share, economic role of tourism, Gini seasonality index, Gini demand structure index, lack of cooperation and environmental impact and a tendency to decline is based on exceeding thresholds of the various indicators. Indicators that require attention to avert a decline are linked to environmental and seasonality issues and also strong excursionist pressure (Manente and Pechlaner, 2006).

Monitoring destination decline is important for effective tourism destination development and requires a broad perspective to include governance structures that facilitate destination development. Interactions of stakeholders are important within a destination since a lack of coordination among interacting tourism stakeholders may result in tourism destination decline. Governance structures should be designed based on the characteristics of a tourism destination network to determine the allocation of resources to a DMO for proper coordination. Such an approach may enhance the destination's competitiveness as the stakeholders with coordination capability will be identified and supported.

10.3 Case Study: The Grand Bahama Island (GBI), the Bahamas

Tourism arrivals in the Bahamas

The Bahamas islands have experienced consistent growth of tourism visitors' numbers over four decades and today the Bahamas is one of the world's leading tourism countries. Its main destinations include Nassau, Paradise Island and the Family Islands. According to the Bahamas Ministry of Tourism, the country comprises 700 islands within an archipelago with a land area of over 100,000 square miles (Bahamas Ministry of Tourism, 2017a). Generally, the country has benefited from its location as the most northern of the Caribbean islands and within close proximity to the US, a significant tourist generating country.

Figure 10.1 shows the growth of visitor arrivals in the Bahamas over four and a half decades. Rapid growth of tourism is noted but there is also evidence of a falloff

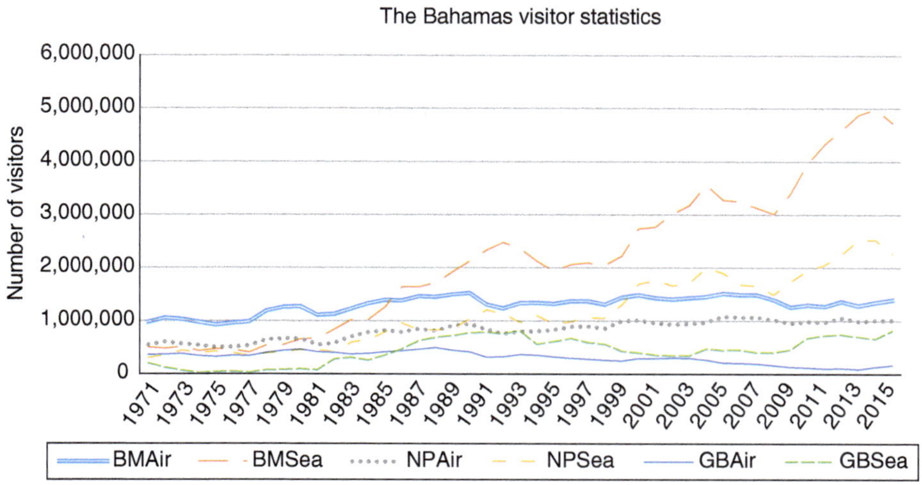

Fig. 10.1. Visitor arrivals in the Bahamas (1971–2015). Key: BM – the Bahamas; NP – New Providence; GB – Grand Bahama Island. (From Bahamas Ministry of Tourism, 2017b.)

in air arrivals to Grand Bahama starting in 1991, with dips in 2004 and 2013 and an improving upward trend since 2014. Within the overall destination, sea arrivals have burgeoned with cruise ship excursionist arrivals far surpassing stay-over visitor arrivals to the islands. Although there was a decrease in sea arrivals between 2004 to 2008, since then there has been consistent growth in sea arrivals unlike that of air arrivals. Within recent years the Grand Bahama Island has experienced declining tourist arrivals, as over the period 2005 to 2013, stopover visitor arrivals declined from 316,262 to 165,602 (Ministry of Tourism, 2014), just about a 90% decline.

The Bahamas and Grand Bahama Island

The city of Freeport is the main tourism destination on the island. Located on the south-west coast of Grand Bahama Island (Fig. 10.2), Freeport is operated under the Hawksbill Creek Agreement (HCA). This agreement commenced on 20 June, 1955 between the government of the Bahamas and a business man, Wallace Groves, and established the city of Freeport and a free trade zone. The HCA states that the governing body should:

> Use their best endeavours to promote and encourage the establishment of factories and other industrial undertakings, and in particular factories, industrial undertakings and industries which will make use of the natural resources and products available at Hawksbill Creek such as limestone rock and pine timber ... (The Government of the Bahamas, 2001, Ch. 261–5).

The agreement granted 50,000 acres of land to the Grand Bahama Port Authority (GBPA) and created in effect an authority to develop and administer the land that functions as a quasi-government. In order to encourage its activities, several tax concessions are allowed. The GBPA built and manages the city of Freeport and has responsibility for the administration and control of the city.

Fig. 10.2. Map of the Bahamas. (From CIA, https://www.cia.gov/library/publications/the-world-factbook/geos/bf.html.)

The Ministry of Tourism and the Grand Bahama Port Authority (GBPA) have shared responsibility for developing tourism in Grand Bahama Island (GBI). The GBPA licensee directory includes a list of 87 tourism operators including aviation, accommodation, attraction, amenities and transport operators (Table 10.1). According to the Bahamas Investor Online (2014) the main plans for developing tourism on GBI include increasing room numbers, increasing airlift capacity as Bahamas Air does not fly from Miami to Freeport, and more cruise line ferry services to the island. The Grand Bahama Island Tourism Board (GBITB) exists to promote the destination in conjunction with the Ministry of Tourism. The Ministry implements the Bahama Host Programme to support product development. In order to provide greater governance of GBI through the central government, a Ministry of Grand Bahama (MGB) was created in 2012 to administer the affairs of the residents of Grand Bahama Island. One goal of the newly created Ministry is to work with owners of closed resorts to reopen them, as following the major hurricanes of 2004 there have been several resorts closures. The Ministry of Grand Bahama notes that a 5-year tax holiday will be given to re-opened properties and a 10-year tax holiday for new properties (Ministry of Grand Bahama, 2017). The creation of MGB further complicates tourism governance on the island., since there are four entities that are delivering DMO activities on GBI; however, the Ministry of Grand Bahama does not have direct responsibility for tourism operators or have tourism operators that are members.

Table 10.1. Tourism operators in the city of Freeport. (From Grand Bahama Port Authority, 2016.)

Type	Number
Aviation	8
Bicycle sales and rental	3
Golf course operation	2
Hotel operation	16
Museum operation	3
Park	2
Tour operation	19
Transportation	4
Water sports	22
Souvenirs	8
Total	87

10.4 Methods

Network theory and tourism destination research

Social network analysis is a method used to map a destination network (see D'Agara, Gozzo and Tomaselli, 2013). Existing relationships within the destination network can be reported based on an interviewee's interactions and those known interactions of others, in other words, data about the destination's network structure formed from reported ties. A tie is a reported relationship between two stakeholders (Borgatti and Halgin, 2011). Freeman, Romney and Freeman (1987) note that informant accuracy can be improved when the 'best' informants report on information regarding long-range stable patterns of events. In this case, the best informants are those senior officials in the DMOs since they interact with all tourism operators on an ongoing basis and organizations with DMO functions were included in this research study. With network analysis, there can be a burden on the informant to collect the number of connections that exist (Borgatti, Everett and Johnson, 2013) and therefore similarly to Beritelli and Bieger (2014) informants were asked to name the ties of up to five other individuals.

Based on the network characteristics in terms of the range of actors and their ties and known ties of others, a network structure is formed. Borgatti and Halgin (2011) propose mechanisms that interact within networks to influence certain outcomes of the networkers. One such outcome that is applicable to destination management is coordination. Coordination is a bonding model based on cooperation and convergence (Borgatti and Halgin, 2011). Cooperation continues through obligations between networkers being met and similarity among actors further create network bonds that support convergence.

Cooperation and convergence occur in the tourism industry as tourism operators become interdependent in the provision of tourism products and services. The cruise lines and hotels cooperate with attraction operators and transport providers. Cooperation means that there is a willingness to share network resources and this builds social capital as tourism operators become obligated to each other. Convergence occurs as similar actors such as hoteliers or attraction operators interact and as the DMO manages tourism products and services. Thus, the tourism and hospitality industries within the destination become coordinated through the interactions of the tourism operators and the DMO within the destination.

Data collection

The overall research design was a network approach that allows for the collection of data from a number of informants about their own interactions and the interactions of others. The data collection method was a semi-structured interview using purposive sampling of the DMOs and main hotel properties.

The research location is one of the islands in the Bahamas with a sufficient number of large hotels to enable the mapping of a tourism destination network. Initially, five main hotel properties were identified by the Ministry of Tourism in Grand Bahama Island and included Memories Blue Diamond Resort, Old Bahama Bay, Pelican Bay Hotel, The Grand Lucayan Resort and Viva Wyndham Fortuna Beach, and these properties were a starting point for data collection. A membership list of nine properties was obtained from the Grand Bahama Island Tourism Board. The General Manager of each property was contacted to be interviewed for the research study. The Ministry of Tourism, the Grand Bahama Port Authority and the Grand Bahama Island Tourism Board were also interviewed. A total of eight questionnaires were obtained from four face-to-face interviews and four telephone interviews.

The research instrument was designed to collect data from informants about their relationships within the destination and the known relationships of those tourism operators that they are connected to. Contextual information was also obtained about the operation of the tourism destination during various time periods. The main sections of the questionnaire were similar to that of Strobl and Peters (2013) in that the instrument included stakeholders' and actors' interactions as well as influence on destination coordination and destination development. Crisis management data were also obtained in relation to how coordination occurs during a crisis. Data were gathered about the activities of accommodation, attraction and transport operators only.

A database of tourism operators managed by the Grand Bahama Port Authority was accessed (Table 10.1) and this was utilized in the destination network mapping using the data from informants. Some operators such as restaurants, cargo brokers and trucking companies were not included in the research study based on two main assumptions: (i) hotel establishments may contain facilities that offer the services of these operators; and (ii) the activity was deemed as secondary as these may not be patronized during a visitor's stay. For businesses with more than one type of operation such as a hotel also operating a water sport activity, only one activity was counted.

Data analysis

The data analysis method was designed to allow the measurement of mapped relationships within a tourism destination network. Once the data were collected the actors were assigned coded identifiers the ties between actors were entered into UCINET 6 software (Borgatti, Everett and Freeman, 2002) and a network pattern revealed using NetDraw software (Borgatti, 2002). The data were symmetrized to show a relationship as symmetrical.

Density and centrality measures were utilized to explore the degree to which actors are interacting to coordinate and influence the growth of the tourism destination. Centrality measures were utilized to understand the influence of the DMOs in the destination. Density is the ratio of actual ties to the total number of ties in the network (McLeod, 2015, p. 149) and central actors are important as they can influence network effectiveness and outcomes (McLeod and McNaughton, 2016, p. 35). Density measures are: (i) size as the number of nodes connected to an actor; (ii) ties as the number of connections in an actor's network; (iii) pairs as the number of possible ties in an actor's network; and (iv) density as the number of ties divided by the number of pairs (Hanneman and Riddle, 2005). Centrality measures according to Freeman (1979) are: (i) degree as the number of actors that an actor is tied to; (ii) closeness as the sum of distances of an actor to all other actors in the network; and (iii) betweenness as the number of times an actor reaches any other actor through another actor (see Borgatti and Everett, 2006).

10.5 Destination Network Management on Grand Bahama Island

Tourist arrivals on the island of Grand Bahama have been declining for more than one decade and an interviewee stated that the main reason for this has been two back-to-back major hurricanes, Frances and Jeanne in 2004 (Interviewees 1 and 6). The island slowly recovered, only to be subject to hurricane Matthew in 2016. Before the 2004 hurricanes the island received significant air connections from major US cities. In 2014 the entry of Sunwing Air based in Canada, owners of the Memories property, resulted in improved stay-over arrivals (Interviewee 1).

Incentives have been provided to boost visitor arrivals. Based on its close proximity to Florida, cruise arrivals are very important. The Grand Celebration with 1,100 to 1,200 passengers and the Balearia Caribbean with 450–600 passengers are two major vessels visiting GBI (Interviewee 1). The average occupancy rate lies between 50–60% per annum and the main tourist origin markets are south Florida, north-east USA and Canada (Interviewees 5 and 8). There is also some domestic traffic through travel for festivals from the other Bahama Islands (Interviewee 1). The Grand Bahama Island provides air credit of US$250 to boost sales once purchase is made through an online travel agency; however, Nassau provides US$500 resort credit, and has cheaper rooms and coupons for spa use (Interviewee 8).

While the Grand Bahama Island offers a mix of activities including diving, water sports, kayaking, fishing, snorkelling, beach ambiance and an ocean environment (Interviewees 4 and 7), it is not viewed as a tourism-dependent destination. One hotelier commented that 'Grand Bahama has a hotel sector that is least dependent on leisure tourism' as 30% of the destination business relates to the shipyard and 25% of business comes from the cruise sailings that offer a 4-nights' package with 2 nights at Grand Lucayan (Interviewee 3). The view is that the destination is 'totally messed up' and a 'roller coaster' as 80% of the rooms belong to the Grand Lucayan strip and half these rooms have been leased to Sunwing (Interviewee 3). Hutchinson Lucaya Limited is an accommodation provider with properties including Lighthouse, Grand Lucayan and Memories, and has 1,300 rooms in total. 'Two years ago the total room revenue was US$18 million' and that amount of revenue is the same as the Dolphin experience in Atlantis, Nassau, the Bahamas (Interviewee 3). Nonetheless, it is the view that timeshare properties are not so affected by the diminished business opportunities (Interviewee, 6).

Destination stakeholders and actors interactions

The interviewees were asked about their interactions with tourism operators in the destination. Two types of data were utilized to construct the destination network of interactions. First, relational data about interviewees own interactions were included. Second, interview data that highlighted known interactions of tourism actors were applied to data collected about the number of tourism operators in the destination using Table 10.1.

The Bahamas Local website was utilized to complete the list of other tourism operators not on the Grand Bahama Port Authority's list of tourism operators (Table 10.1). Grand Bahama Island stakeholder interactions were identified based on snowball sampling beginning with the views from staff of Destination Management Organizations (Table 10.2). Other categories of businesses such as

Table 10.2. Grand Bahama Island stakeholders (interviewees' known interactions).

Airlines interact with travel agents, cargo brokers, trucking, business houses and food stores	Cruise lines interact with travel agents, cargo brokers, trucking, business houses and food stores
Grand Bahama Port Authority interacts with accommodation providers, straw vendors, attractions, restaurants, investors	Grand Bahama Airport Company Limited interacts with Grand Bahama Port Authority, private airport in West End
All accommodation operators interact with Ministry of Tourism, Grand Bahama Tourism Board, airlines and cruise lines	Attractions interact with cruise lines, tour operators
Freeport Bahamas Adventures interacts with Carnival cruise lines, Bahamas Celebration, Grand Lucayan hotel, Memories, Pelican Bay and Balearia Caribbean	H. Forbes Charters interacts with Carnival cruise lines, Bahamas Celebration, Grand Lucayan hotel, Memories, Pelican Bay and Balearia Caribbean
Sunwing interact with Ministry of Tourism and UNEXSO	Jeep Safari tours interact with all hotels, cruise ship and boating charters
Grand Bahama Island Tourism Board interacts with all stakeholders	

business houses, restaurants and investors and non-tourism oriented businesses such as cargo brokers and trucking businesses were not included.

Figure 10.3 illustrates the tourism destination network with three central organizations: Ministry of Tourism (DMO1), Grand Bahama Port Authority (DMO2) and the Grand Bahama Island Tourism Board (DMO3). DMO2 has the highest density value of 5.01% (242 / 4830 × 100) (Table 10.3). On one hand, the highest density of DMO2 gives Grand Bahama Port Authority the best coordination capability. On the other hand, DMO3 has the highest degree centrality, highest betweenness and is located close to DMO1. DMO2 is least close to all other nodes. Based on its centrality, DMO3 seems in the best position to influence the management and development of this tourism destination. The importance of centrality relates to where actors are located as actors within close proximity to other actors can influence the sharing of resources.

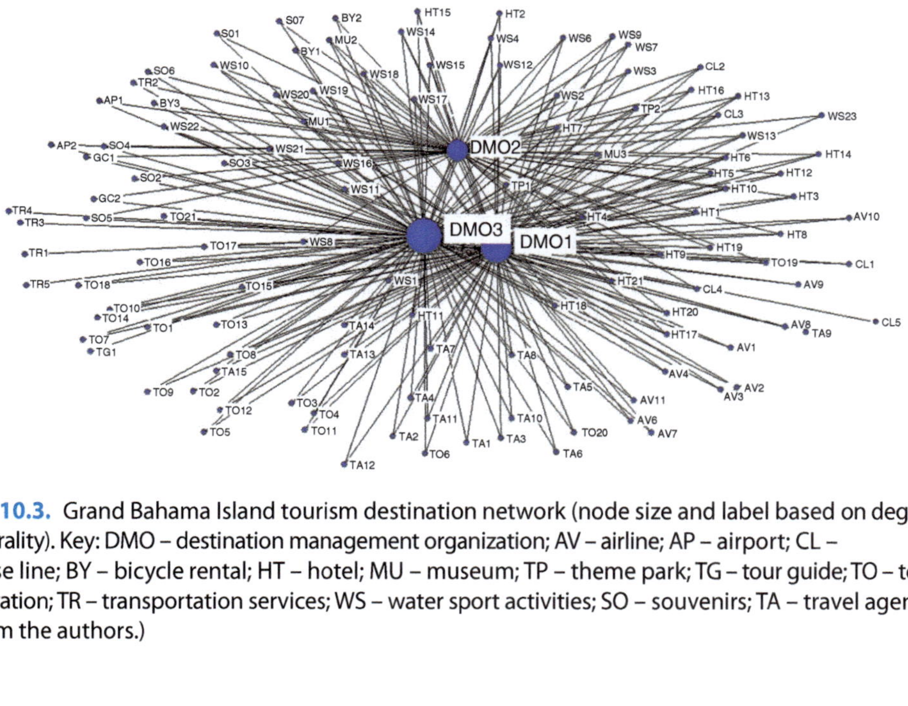

Fig. 10.3. Grand Bahama Island tourism destination network (node size and label based on degree centrality). Key: DMO – destination management organization; AV – airline; AP – airport; CL – cruise line; BY – bicycle rental; HT – hotel; MU – museum; TP – theme park; TG – tour guide; TO – tour operation; TR – transportation services; WS – water sport activities; SO – souvenirs; TA – travel agent. (From the authors.)

Table 10.3. Grand Bahama Island tourism destination network (density and centrality).

Destination manager	Density measures				Centrality measures		
	Size	**Ties**	**Pairs**	**Density**	**Degree centrality**	**Closeness centrality**	**Betweenness centrality**
DMO1	103	306	10,506	2.91	103	143	2,426.0
DMO2	70	242	4,830	5.01	70	176	943.5
DMO3	121	334	14,520	2.30	121	125	3,990.5

Destination coordination and crisis management

Three organizations on Grand Bahama Island deliver destination management functions: the Ministry of Tourism, the Grand Bahama Port Authority and the Grand Bahama Island Tourism Board (GBITB). The role of the Ministry of Tourism involves 'creating, maintain and improving the tourism product' (Interviewee 1). The GBPA is responsible for issuing licenses to tourism operators in the city of Freeport and the Ministry of Tourism issues licenses to tourism operators in the other parts of Grand Bahama Island (Interviewee 1). The GBITB performs joint marketing of the destination.

Freeport is a private city managed by the Grand Bahama Port Authority based on the proclamation of the Hawksbill Creek Agreement (Interviewee 1). GBPA is responsible for business approvals and amendments to business licenses. The main functions of the Authority include:

- business license approval;
- building and sanitary code compliance;
- town planning regulation;
- environmental compliance;
- city management;
- utility regulation; and
- economic development: medical tourism, education tourism and religious tourism as well as other growth sectors (Interviewee 2).

Hurricanes are a major threat to sustainability of tourism on the island. The Ministry of Tourism completes a damage assessment exercise after a hurricane to assess the physical state of the plant. All stakeholders are advised to 'speak on same line' when discussing a crisis situation (Interviewee 1). During a hurricane there are various states of alert and NEMA (National Emergency Management Agency) is involved. A weather conference has been held for the past 5 or 6 years to educate US and Canadian meteorologists about the Bahamas so that if a hurricane is impacting the southern part of the Bahamas this does not mean that the entire country is being affected. Emergency teams come together during a hurricane event and a guest population count on each island is taken so that hotel guests can be evacuated if necessary. Guests are asked to sign a waiver if not evacuated and they go to a shelter, then they are subject to the same requirements as residents. There are shelters throughout the islands. There are local government meetings to plan for crises and crisis management drills for any emergency (Interviewee 1).

Destination stakeholders form partnerships to ensure smooth operations of the destination during a crisis. A hotel hosts the first response group and the US pre-clearance at the airport moves into a hotel (Interviewee 3). One hotel mentioned that there is a hurricane warehouse that stocks water, canned food and an internal generator to support a disaster (Interviewee 4).

Destination development

The Ministry of Tourism plays a key role in hotel and airlift development and cruise/ferry development (Interviewee 5) and also advertising and promoting the destination

(Interviewee 7). Standards for tourism operators are managed by the Ministry of Tourism and there are several committees that manage the destination (Interviewee 1). The ground transport committee manages concerns within transport and includes the Ministry with responsibility for road traffic. The safety and security committee has officers in hotels and manages the beaches. There is a Police Tourism Unit that is particularly active during spring break. The Ministry of Tourism works along with Bahamian owners and stakeholders to develop authentic souvenirs, seeks to ensure that residents are the main benefactors of tourism business, and that superior customer service is provided. There is emphasis on the way to interact with guests such as through the 'People to People' programme (Interviewee 1).

Intercontinental Diversified Corporation is the organization controlling the Grand Bahama Port Authority and the Port Group Limited. Under the Port Group Limited are several companies including Grand Bahama Development Company (50% ownership); Water Utility (100% ownership); Airport Company (50% ownership) and Harbour Company (50% ownership). The Grand Bahama Development company is responsible for hotel development (Interviewee 2). The power generation is owned by the Grand Bahama Port Authority and the cost of electricity is much higher as a result (Interviewee 1).

Grand Bahama Port Authority (GBPA) conducts business development for tourism and other business sectors and encourages hotel investment. A small business bureau has also been established to handle investment promotions for Grand Bahama Island, called Invest Grand Bahama Small Business Bureau. The Bureau also provides capacity building for local businesses through education and training (Interviewee 2). Also, the GBPA conducted a feasibility study regarding medical tourism in 2009. The study recommended that the destination is suitable for low-risk, high-volume procedures such as orthopaedics, cosmetic surgery and stem cell surgery. In 2013, the government passed legislation to facilitate the development of medical tourism. The Okyanos Heart Stem Cell Institute was attracted to Grand Bahama. Education tourism has also been introduced to attract Executive Education programmes to the destination. University of Pennsylvania, Wharton Business School and other universities host their programmes on the island. There are religious conferences and international ministries, and work was being done with the late Dr Myles Munroe to develop religious conferences (Interviewee 2).

The Grand Bahama Port Authority works with the Bahamian government to grow and develop the sector. Nonetheless, there has not been 'a new hotel development in 10–12 years, hotels close and do not re-open' (Interviewee 5). This is attributed to costs of financing, challenges with the infrastructure and difficulty to transport visitors to the other side of the island (Interviewee 7). Businesses also experience high fuel and labour costs. Another view is that government has investments in Nassau, for example the Baha Mar Hotel, and is not willing to do anything on the Grand Bahama Island (Interviewee 6). Several business operators seek to ensure repeat business through enhancing the guest experience (Interviewee 8). One hotelier plans a major renovation of US$5 million to improve its product (Interviewee 4). Grand Bahama Island has a privately owned airport with the highest fees in the region and therefore the island does not attract low-cost carriers (Interviewee 1). With no low-cost carrier (LCC) calling at the destination, the airfares are higher than going to Nassau although the distance is shorter (Interviewee 5).

10.6 Conclusion and Lessons Learned

Closure of hotel rooms and withdrawal of airlines are key signs of a destination in decline, a consequence is the remaining hotels engage in intense competition that in turn limits cooperative interactions among the various stakeholders within the destination. This chapter has analysed the tourism destination network to identify key stakeholders and used interviews with stakeholders to identify destination management practices. The network analysis indicates that there are three central DMOs and the stakeholder interviews indicates that these organizations do not cooperate and have different objectives with regard to tourism development. These problems are compounded by the destination's susceptibility to natural disasters. Hurricane Matthew brought tourism activities to a halt, but many properties assessed damage, reopening a couple of months after the hurricane and retrained staff during the down period (Bahamas Ministry of Tourism, 2016). However, this recovery was at a significant reduction in profitability.

This case provides an opportunity to explore the role of DMOs within the destination. The three organizations that have capability to coordinate and influence the tourism destination network were identified. The results suggest a need to examine each stakeholder's role within the destination in order to improve destination management practices. Greater coordination is required to create sustained tourism performance, especially when a destination is recovering from a crisis. The smooth performance of tourism activities depends not only on the actor, but also the role this actor undertakes in coordinating tourism destination activities. DMO2 for instance has an important role in the tourism destination network; however, this organization does not have a leading role in marketing the destination and seems to be constrained in destination development activities such as attracting new hotel investment. Based on its structure, the GBPA is not a tourism-oriented organization and its capacity to manage and develop tourism products and services could strengthen through building knowledge and allocating resources towards tourism as a main coordinating organization. Without a change in direction of the role played by DMO2 and certain actions taken such as a rethink of airport fees, tourism destination growth will be stymied. Analysis of a tourism destination network supports destination development and may help to avert a sustained decline in tourism destination performance.

An effective DMO is important to coordinate stakeholders and address destination decline. Clearly there are issues with the various DMOs being able to effectively coordinate, contributing in part to a decline in tourism on Grand Bahama Island. The network analysis shows three DMO organizations that should review their roles to ensure that policies are instituted to improve tourism destination performance. Interestingly, the DMO that is most central is least dense, DMO3. Although DMO3 has access to stakeholders, the number of ties with stakeholders suggest a lack of coordination. This analysis indicates the value of analysing destination networks as it identifies the most central DMO, their stakeholder ties, and thereby coordination, so that policies that allow the network to operate more effectively can be implemented.

This research was conducted in a destination that is in decline. Future research can be conducted in other destinations at the different stages of the resort life cycle (Butler, 1980) to compare and contrast destination networks. The various roles played by DMOs can be another avenue of future research particularly the role of a DMO in islands that are developing other industries.

Acknowledgements

The authors extend appreciation to Mrs Edith Thompson of Grand Bahama Island for her hosting of the researcher during the data collection for this chapter.

References

Baggio, R. and Sainaghi, R. (2016) Mapping time series into networks as a tool to assess the complex dynamics of tourism systems. *Tourism Management* 54, 23–33.

Bahamas Local (2017) Available at:. https://www.bahamaslocal.com/ (accessed 26 January 2017).

Bahamas Ministry of Tourism (2016) Grand Bahama tourism industry begins rebuilding after Hurricane Matthew Available at: http://www.bahamas.com/article/grand-bahama-tourism-industry-begins-rebuilding-after-hurricane-matthew (accessed 24 January 2017).

Bahamas Ministry of Tourism (2017a) The Islands of the Bahamas. http://www.bahamas.com/explore-our-islands, Accessed 24th January, 2017.

Bahamas Ministry of Tourism (2017b). Tourism Today Network. Available at: http://www.tourismtoday.com/sites/default/files/air_sea_arrivals_71_thru_2015_0.pdf, (accessed 15 February 2017).

Becken, S., Scott, N. and Ritchie, B.W. (2015) The development of new tourism networks to respond to and recover from the 2011 Christchurch earthquake. In: Ritchie, B.W. and Campiranon, K. (eds) *Tourism Crisis and Disaster Management in the Asia–Pacific*. CAB Introduction, Wallingford, UK, pp. 190–205.

Beritelli, P. and Bieger, T. (2014) From destination governance to destination leadership – defining and exploring the significance with the help of a systemic perspective. *Tourism Review* 69(1), 25–46.

Borgatti, S.P. (2002) *Netdraw Network Visualisation*. Analytic Technologies, Harvard, Massachusetts.

Borgatti, S.P. and Everett, M.G. (2006) A graph-theoretic perspective on centrality. *Social Networks* 28(4), 466–484.

Borgatti, S.P., and Halgin, D. (2011) On network theory. *Organization Science* 22(5), 1168–1181.

Borgatti, S.P., Everett, M.G. and Freeman, L.C. (2002) *UCINET for Windows: Software for Social Network Analysis*. Analytic Technologies, Harvard, Massachusetts.

Borgatti, S.P., Everett, M.G. and Johnson, J.C. (2013) *Analyzing Social Networks*. SAGE Publications Ltd., London.

Butler, R. (1980) The concept of a tourist area cycle of evolution: implications for management of resources. *Canadian Geographer* 24(1), 5–12.

Butler, R. (2006) *The Tourism Area Life Cycle: Conceptual and Theoretical Issues*, Vol. 2. Channel View Publications, Clevedon, UK.

Central Intelligence Agency (2017) The World Factbook, Central America and Caribbean: Bahamas, The. Available at: https://www.cia.gov/library/publications/the-world-factbook/geos/bf.html (accessed 9 March 2017).

D'Agata, R., Gozzo, S. and Tomaselli, V. (2013) Network analysis approach to map tourism mobility. *Quality and Quantity* 47(6), 3167–3184.

Diedrich, A. and Garcia-Buades, E. (2009) Local perceptions of tourism as indicators of destination decline. *Tourism Management* 30(4), 512–521.

Freeman, L.C. (1979) Centrality in networks: conceptual clarification. *Social Networks* 1(3), 215–239.

Freeman, L.C., Romney, A.K. and Freeman, S. (1987) Cognitive structure and informant accuracy. *American Anthropologist* 89(2), 310–325.

Grand Bahama Port Authority (2016) Licensee Directory. GPPA, Grand Bahama Island.

Hanneman, R. and Riddle, M. (2005) *Introduction to social network methods*. University of California, Riverside, California.

Hristov, D. and Zehrer, A. (2015) The destination paradigm continuum revisited: DMOs serving as leadership networks. *Tourism Review* 70(2), 116–131.

Komppula, R. (2014) The role of individual entrepreneurs in the development of competitiveness for a rural tourism destination – A case study. *Tourism Management* 40, 361–371.

Laesser, C. and Beritelli, P. (2013) St. Gallen consensus on destination management. *Journal of Destination Marketing and Management* 2(1), 46–49.

Laws, E. (1995) *Tourist Destination Management: Issues, Analysis and Policies.* Routledge, London.

Lew, A. and McKercher, B. (2006) Modeling tourist movements: a local destination analysis. *Annals of Tourism Research* 33(2), 403–423.

Manente, M. and Pechlaner, H. (2006) How to define, identify and monitor the decline of tourist destinations: towards an early warning system. In: Butler, R. (ed.) *The Tourism Area Life Cycle: Conceptual and Theoretical Issues, Vol. 2.* Channel View Publications, Clevedeon, UK, pp. 235–253.

McLeod, M. (2015) Analysing inter-business knowledge sharing in the tourism sector. In: McLeod, M. and Vaughan, R. (eds) *Knowledge Networks and Tourism.* Routledge, Oxford, UK, pp. 143–157.

McLeod, M. and McNaughton, M. (2016) Mapping an emergent Open Data eco-system. *Journal of Community Informatics* 12(2), 26–46.

Merinero-Rodriguez, R. and Pulido-Fernández, J.I. (2016) Analysing relationships in tourism: A review. *Tourism Management* 54, 122–135.

Ministry of Grand Bahama (2017) The revitalization and promotion of Grand Bahama as a tourism and industrial centre. The Government of the Bahamas, the Bahamas. http://www.bahamas.gov.bs Accessed 16th February, 2017.

Ministry of Tourism (2014) Frequently requested tourism statistics. Research and Statistics Department, Ministry of Tourism, the Bahamas.

Morakabati, Y., Page, S.J. and Fletcher, J. (2016) Emergency management and tourism stakeholders' responses to crises a global survey. *Journal of Travel Research* doi: 10.1177/0047287516641516.

Ness, H., Aarstad, J., Haugland, S.A. and Grønseth, B.O. (2014) Destination development: the role of interdestination bridge ties. *Journal of Travel Research* 53(2), 183–195.

Pavlovich, K. (2014) A rhizomic approach to tourism destination evolution and transformation. *Tourism Management* 41, 1–8.

Pearce, D.G. (2014) Toward an integrative conceptual framework of destinations. *Journal of Travel Research* 53(2), 141–153.

Pearce, D.G. (2015) Destination management in New Zealand: Structures and functions. *Journal of Destination Marketing & Management* 4(1), 1–12.

Pforr, C., Pechlaner, H., Volgger, M. and Thompson, G. (2014) Overcoming the limits to change and adapting to future challenges governing the transformation of destination networks in Western Australia. *Journal of Travel Research* 53(6), 760–777.

Ritchie, J.R.B. and Crouch, G.I. (2003) *The competitive destination: a sustainable tourism perspective.* CAB International, Wallingford, UK.

Scott, N., Cooper, C. and Baggio, R. (2008) Destination networks: four Australian cases. *Annals of Tourism Research* 35(1), 169–188.

Strobl, A. and Peters, M. (2013) Entrepreneurial reputation in destination networks. *Annals of Tourism Research* 40, 59–82.

The Bahamas Investor (2014) New strategies in tourism. Available at: http://www.thebahamasinvestor.com/2014/new-strategies-in-tourism/ (accessed 26 January 2017).

The Government of the Bahamas (2001) Hawksbill Creek, Grand Bahama (Deep Water Harbour and Industrial Area). Statute Law of the Bahamas Agreement, the Bahamas.

Volgger, M. and Pechlaner, H. (2014) Requirements for destination management organizations in destination governance: understanding DMO success. *Tourism Management* 41, 64–75.

World Tourism Organization (2007) A Practical Guide to Tourism Destination Management. World Tourism Organization, Madrid.

11 Music Festivals in Small Island Destinations

Kelly J. Semrad,* Manuel Rivera and Robertico Croes

University of Central Florida, United States of America

11.1 Introduction

Since the 1990s, festival production has seemingly increased as the world's consumption patterns drifted from purchasing production goods and services to the experience economy, whereupon consumers searched for experiential products in which they could immerse themselves during the consumption process (Semrad and Rivera, 2016). Festivals are a form of special events that may be considered one of the most experiential event products to occur in modern event production. Generally speaking, festivals represent a series of organized acts, performances or a themed period of time set aside for feasting and celebration. Festivals are often staged by a community and are frequently produced to celebrate a unique facet, interest or cause shared by the members of that community. Note that community is not necessarily defined as the location of where people live, but could also be defined as a reference or social group that an individual subscribes to or belongs. Thus, there are a wide array of different types of festivals that are used to reflect such community interests and causes such as religious, art, food, agricultural, wine, beer, sports, cultural, seasonal, film, dance, folk, comedy, lifestyle, athletic, political, health, music, amongst a plethora of other types of festivals.

Festival production has also infiltrated communities that share a common economic interest or benefit from a particular industry, such as tourism. Andersson and Getz (2009) classify festivals as experiential products that may be offered as tourism attractions in order to draw local, domestic and international event attendees to a destination during a specific time period. Tourism destinations frequently use festivals as a means of event tourism to promote community interest and exposure that may lead to increased tourist arrivals during a low-demand period, thus stimulating the local and/or regional economy (Felsenstein and Fleisher, 2003). The benefit of destinations using festivals as experiential attractions within the tourism product portfolio is that the attraction is finite and temporary. This means that festivals do not require

*E-mail: kelly.semrad@ucf.edu

permanent space or resources and may constantly be 'refreshed' or updated to appeal to either different market segments that frequent a destination, or market segments that a destination is attempting to acquire – all the while still diversifying the tourism product portfolio with different experiences.

Destinations' use of festivals as experiential tourism attractions has led to five distinct areas of academic research. According to Kim *et al.* (2010) the five primary research streams that have emerged due to destinations' positioning festivals as tourism attractions include the following: (i) attendees' motivations for consuming festivals; (ii) residents' and/or attendees' perceptions of festivals; (iii) economic impacts of festivals; (iv) methods used to assess economic impacts of festivals; and (v) characteristics that influence consumption patterns of festival attendees. Table 11.1 provides several academic literature references that studied at least one of these five research streams.

11.2 Music Festivals

Amongst the different types of festivals used to stimulate tourism, music festivals have emerged as one of the most popular forms of festival production thus resulting in most tourist destinations, cities and regions integrating various genres of music festivals as attractions within a destination's tourism portfolio (Frey, 1994; Rivera *et al.*, 2016). Morgan (2008) claims that this increase in music festival consumption may be due to consumers' tastes and preferences for live music that is consumed in an immersive environment that engages festival attendees with a stimulating event atmosphere that includes other people and music. In the tourism context, consumption of a music festival becomes an even more immersive environmental experience due to the event atmosphere encompassing the destination's culture and foreign backdrop of the location of the festival.

Small island destination (SID) music festivals

According to Semrad and Rivera (2016), a new, yet consistent, emerging music festival marketplace is that of island destinations that are located within the Caribbean region. Island destinations located in appealing climates seem to provide one of the most desirable forms of environments for tourists to immerse themselves within, as they are able to enjoy tropical weather, the local island culture, live music as an experiential product, as well as other tourists and locals at the festival. Table 11.2 provides a list of some of the better known music festivals that occur in the Caribbean region.

The growing popularity of Caribbean island destinations using music festivals as a means to boost tourism during low-demand seasons seems especially forthcoming in small island destinations (SIDs), whereupon the top ten beneficiaries of tourism on a worldwide scale are SIDs (Schubert *et al.*, 2011). This is because SIDs possess some prevalent destination attributes that appeal to the tourist masses by way of a triple S product (i.e. sun, sand and sea). However, these destination attributes (triple S product) are not necessarily enough to provide SIDs with a competitive advantage against other destinations that possess the same or similar attributes of the sun, sand and sea (Semrad and Rivera, 2015).

Table 11.1. Five research themes of festivals as tourism attractions.

Festival research stream	Title of article and journal	Author(s) and date
Attendees' motivations	A review of festival & event motivation studies. *Event Management* 9(4).	Li, X. and Petrick, J. (2006)
Attendees' motivations	Motives of visitors attending festival events. *Annals of Tourism Research* 24(2).	Crompton, J. and McKay, S. (1997)
Residents' and/or attendees' perceptions of festivals	Perceived impacts of festivals and special events by organizers: An extension and validation. *Tourism Management* 25(2), 171–181.	Gursoy, D., Kim, K. and Uysal, M. (2004)
Residents' and/or attendees' perceptions of festivals	Development of a scale to measure resident attitudes toward the social impacts of community festivals, Part II. Verification of the scale. *Event Management* 7(1).	Delamare, T. (2001)
Economic impact of festivals	The internationalization benefits of music festivals: The case of the Curacao North Sea Jazz Festival. *Tourism Economics* 22(5).	Rivera, M., Semrad, K. and Croes, R. (2016)
Economic impact of festivals	Festival tourism: A contributor to sustainable local economic development? *Journal of Sustainable Tourism* 10(4).	O'Sullivan, D. and Jackson, M. (2002)
Methods to assess economic impact of festivals	A methodology for estimating the attendance and economic impact of an open-gate festival. *Event Management* 9(3).	Tyrell, B. and Ismail, J. (2005)
Methods to assess economic impact of festivals	A guide for undertaking economic impact studies: The Springfest example. *Journal of Travel Research* 40(1).	Crompton, J., Lee, S. and Shuster, T. (2001)
Factors influencing attendees' expenditure patterns	The economic impact of rural festivals and special events: Assessing the spatial distribution of expenditures. *Journal of Travel Research* 28(4).	Long, P. and Perdue, R. (1990)
Factors influencing attendees' expenditure patterns	A comparison of results of three statistical methods to understand the determinants of festival participants' expenditures. *International Journal of Hospitality Management*, 29(1).	Kim, S., Prideaux, B. and Chon, K. (2010)

The academic references above were compiled by the authors. The articles in the list are just a few of the publications that pertain to the top five research streams within the realm of festival management for tourism destinations.

Thus, Prentice and Andersen (2003) and Richards and Wilson (2005) suggest that destinations, such as SIDs, may work to distinguish themselves apart from other triple S (SSS) destinations by way of creative tourism, which incorporates efforts to diversify the tourist experience with experiential attractions. Music festivals are one way that SIDs may incorporate the use of creative tourism by adding experiential tourist attractions to the destination portfolio without permanent investment or

Table 11.2. Caribbean music festivals. (From Semrad and Rivera, 2016.)

Festival name	Country	Festival name	Country
Soul Beach Music Festival	Aruba	The Red Strip Sumfest	Jamaica
Aruba Piano Festival	Aruba	Rebel Salute Music Festival	Jamaica
Jazz and Latin Music Festival	Aruba	Ocho Rio Jazz Festival	Jamaica
Junkanoo Festival	Bahamas	Jamaica Rock and Blues Jam	Jamaica
Barbados Jazz Festival	Barbados	Reggae Sumfest	Jamaica
Barbados Gospelfest	Barbados	St. Lucia Jazz Festival	St. Lucia
Bermuda Music Festival	Bermuda	Rhythm and Blues Festival	US Virgin Islands
Curacao North Sea Jazz	Curacao	Tobago Jazz Experience	Trinidad and Tobago
St. Kitts Music Festival	St. Kitts	Jazz Artists on the Green	Trinidad and Tobago

allocation of resources. As previously referenced, this is because music festivals are temporary in nature. In other words, the event is planned, produced, executed and is concluded until the next year's production. All the while the destination may continue to reap the economic and marketing benefits of a well-produced music festival.

The benefit of a temporary attraction that occurs in a SID and falls within the realm of creative tourism, such as a music festival, is that there is limited space (land) available on an island, so any space allotted for a tourism attraction must have longevity in the market place. This means that the attraction must appeal to tourists over the course of time; otherwise the space (land) could have been better used by an attraction that would continuously generate tourist interest to arrive to the SID and continuously generate revenue for the island's economy over time. The problem for most tourist attractions (regardless of location: mainland or island destination) is that consumer tastes and preferences undoubtedly and continuously change over time. Thus, in the case of a music festival occurring in a SID, the space allocated for the event production is not permanent and the music festival can be changed, continuously refreshed or even discontinued to accommodate a destination's needs to provide attractions that are more attractive to tourists' changing tastes and preferences.

The benefits of production and execution of music festivals occurring in SIDs may extend beyond the diversification of island tourist experiential attractions to also assist destination managers with overcoming some unique market challenges that confined economies such as SIDs may face. Srinivasan (1986) and Easterly and Kraay (1999) reference several market compositional challenges that small economies face when compared to that of larger economies such as: lack of economies of scale, market vulnerability, remoteness, decreased access to capital markets, macroeconomic policy dependence and overstatement of real income. These compositional challenges coupled with governance and other latent factors result in most SIDs falling within a stead of perpetual developing economies. However, the assembly of a SID music festival that induces the market demand necessary to increase tourist arrivals and thus foreign exchange via the use of the local tourism supply chain may act as a driver to stimulate the tourism industry to serve as a force that boosts the larger economy. For example, Semrad and Bartels (2014) claim that a confined economy that actively engages its economic linkages to support the production of tourism products and attractions will reap economic rewards for the tourism supply chain thus providing financial stimulation to the local economy.

The current chapter reviews the empirical results from eight different SID music festival productions (six productions of the Curacao North Sea Jazz Festival and two productions of the Aruba Soul Beach Music Festival). The empirical results are themed according to the aforementioned five general research streams in festival management literature: that is, attendees' motivations, residents' and/or attendees' perceptions of the festival, the economic value of festivals, methods used to assess the economic value of the festivals and factors influencing attendees' expenditure patterns (Kim *et al.*, 2010). SIDs are defined as island destinations with a total population of one million or less (Croes, 2011).

This chapter references Curacao and Aruba as specific Caribbean SID cases to highlight the potential economic benefits that music festivals may bring to confined economies. This chapter shares longitudinal research findings from the Curacao North Sea Jazz Festival and the Aruba Soul Beach Music Festival spanning the time frame of 2010–2015, as well as relevant literature that may be used to help support or explain the empirical findings from the aforementioned cases.

11.3 The Cases: Curacao North Sea Jazz Festival and Aruba Soul Beach Music Festival

The Curacao North Sea Jazz Festival is an annual music festival that occurs in the beginning of September in Willemstad, Curacao. The festival is a Caribbean rendition of the well-known North Sea Jazz Festival that occurs in Rotterdam each year. The North Sea Jazz Festival has been in production in the Netherlands since 1976 and is the largest indoor music festival in the world. The Curacao North Sea Jazz Festival has been in production since 2010 and has grown steadily each year since it has been in production boasting significant economic impacts to the island.

The Aruba Soul Music Festival is an annual six-day music festival that occurs Memorial Day weekend (end of May) in the port of Aruba (Oranjestad). The festival has been in production for nearly 20 years and features a concert series, daytime events, nightclub parties and comedy events. The festival was voted one of the 'Top Five Caribbean Celebrations' by USA Today, as well as the 'Number one Caribbean Summer Festival' by Jetsetter Magazine. Both the Curacao North Sea Jazz Festival and the Aruba Soul Beach Music Festival are well known for their beautiful tropical island backdrop and famous artist lineups. For more information regarding these music festivals please visit: http://www.curacaonorthseajazz.com/nl/ and/or http://soulbeach.net/2017/

Although Curacao and Aruba are in different destination life cycle phases (i.e. Aruba is mature and Curacao is in a brand awareness phase), the destinations both use the music festivals in a similar manner, which is to attract and increase tourist arrivals to the island destinations during a specific period of time. Thus, for both of these music festivals it was assumed during the statistical analyses, and later empirically validated, that there were two prominent target audiences that will attend these music festivals: local residents and/or tourists. This is because of the confined nature of a SID economy whereupon one target audience (residents) lives on the island and any other festival attendee must arrive by travel to the island destination in order to consume the live event (tourists). Thus, due to the high volume of tourist arrivals the festivals have been largely deemed responsible for generating an increase in foreign exchange during the time the events occurred.

Table 11.3 provides the themed results from each of the eight festival productions that were examined (2010–2015). A brief description of the results for each annual festival production follows Table 11.3.

11.4 Longitudinal Research Findings

The current chapter provides the longitudinal research findings primarily for the Curacao North Sea Jazz Festival. These findings constitute six years of data analysis that includes the music festival's inception year (2010) all the way through to its fifth year of production (2015). The data provides a relatively robust analysis of the festival's growth and its economic impact to the island of Curacao. The chapter also includes reference to the Aruba Soul Beach Music Festival. Although only two years of data assessment were performed for the Soul Beach Music Festival the data revealed some interesting findings pertaining to a new market that was frequenting Aruba in order to attend the music festival.

Aruba Soul Beach Music Festival (years 2011 and 2012)

2011 The economic and tourism potentials of the Aruba Soul Beach Music Festival

In the report entitled *The Economic & Tourism Potentials of the Aruba Soul Beach Music Festival* (Croes *et al.*, 2011), it was revealed that the music festival was successful from a variety of perspectives. The festival provided the island of Aruba with important economic benefits via the spending of the festival production itself, as well as its tourist attendees. According to the festival assessment, the majority of the festival attendees were female from the Generation X, highly educated, with high income (i.e. five out of ten of these tourists had an income that exceeded $100,000) and most of them had not been to Aruba in the past (54% were first-time arrivals to the island).

Seven out of ten international festival attendees indicated that their main purpose for coming to Aruba was to attend the music festival. The data assessment seemed to indicate that the festival served as a unique vacation experience for an extended stay 'girls night out'. The benefit here is that this market segment (Gen X, affluent and educated women) varies substantially from Aruba's traditional tourist, who is a senior, less affluent and usually a repeat tourist. Thus, the music festival in 2011 attracted a new market segment to the island of Aruba and generated an economic impact of $1.8 million with a multiplier effect of 2.31. The festival also eased seasonality for the month of May and marketed Aruba by showcasing a positive tourism experience.

2012 The Aruba Soul Beach Music Festival: getting the fundamentals right

In the report entitled *The Aruba Soul Beach Music Festival: Getting the Fundamentals Right* (Croes and Rivera, 2012), the data assessment confirmed the results from the

Table 11.3. Themed research results from festival assessments 2010–2015.

Festival	Report name	Major findings	Research themes	Website address
2011 Aruba Soul Beach Music Festival	The Economic & Tourism Potentials of the Aruba Soul Beach Music Festival	(i) Festival attracted a different topography of tourists (ii) Attendees were satisfied with the event (iii) Festival made a positive economic impact	(i) Attendees' perceptions of the music festival (ii) Economic value of the music festival	Web address not available
2012 Aruba Soul Beach Music Festival	The Aruba Soul Beach Music Festival: Getting the Fundamentals Right	(i) Confirmed the findings from the 2011 festival production	(i) Attendees' perceptions of the music festival (ii) Economic value of the music festival	Web address not available
2010 Curacao North Sea Jazz Festival	Uncovering the Potential of a New More Affluent and Loyal Market: The Curacao North Sea Jazz Festival	(i) Majority of attendees indicated the festival exceeded expectations (ii) Festival attracted a different topography of tourists (iii) Festival made a positive economic contribution	(i) Attendees' perceived value of the music festival (ii) Economic value of the music festival	http://stars.library.ucf.edu/dickpope-pubs/14/
2011 Curacao North Sea Jazz Festival	The Signature Event in Curacao: A Source of Brand Equity and Economic Significance	(i) Majority of attendees indicated the festival exceeded expectations (ii) Festival attracted a different topography of tourists (iii) Festival made a positive economic contribution (iv) Festival beginning to establish a loyal following	(i) Attendees' perceived value of the music festival (ii) Economic value of the music festival	http://stars.library.ucf.edu/dickpope-pubs/12/

Continued

Table 11.3. Continued.

Festival	Report name	Major findings	Research themes	Website address
2012 Curacao North Sea Jazz Festival	Exceeding Expectations & Evoking Synergies for Future Growth	(i) Festival continues to exceed expectations, tourists coming for multiple days of the event (ii) Festival attracted a different topography of tourists (iii) Festival made a positive economic contribution (iv) Festival continuing to establish a loyal following and is deemed memorable	(i) Attendees' perceived value of the music festival (ii) Economic value of the music festival	http://stars.library.ucf.edu/dickpope-pubs/1/
2013 Curacao North Sea Jazz Festival	The Curacao North Sea Jazz Festival: Through the Looking Glass	(i) Dutch economic recession impacts attendance rates (ii) Festival continues to make a positive economic impact (iii) Loyalty to the festival may be reaching a saturation point	(i) Attendees' perceived value of the music festival (ii) Economic value of the music festival	http://stars.library.ucf.edu/dickpope-pubs/18/?utm_source=stars.library.ucf.edu%2Fdickpope-pubs%2F18&utm_medium=PDF&utm_campaign=PDFCoverPages
2014 Curacao North Sea Jazz Festival	The Curacao North Sea Jazz Festival: A Lustrum Churning Economic and Promotional Opportunities for Curacao	(i) Festival continues to make a positive economic contribution (ii) Festival helps to enhance the image of the destination (iii) Festival has seemingly developed a brand name apart from the destination	(i) Attendees' perceived value of the music festival (ii) Economic value of the music festival	http://stars.library.ucf.edu/dickpope-pubs/37/?utm_source=stars.library.ucf
2015 Curacao North Sea Jazz Festival	Curacao North Sea Jazz & Destination Convergence: A Harbinger Beckoning?	(i) Satisfaction and likelihood to recommend music festival are high (ii) Festival continues to make a positive economic contribution (iii) Attendees continue to grow more loyal to the festival	(i) Attendees' perceived value of the music festival (ii) Economic value of the music festival	http://stars.library.ucf.edu/dickpope-pubs/39/

2011 festival production. The 2012 rendition of the music festival revealed that the affluent Gen X females who were first-time visitors to the island were still a primary audience at the music festival. In the 2012 assessment, a secondary new target audience attended the event. This new target audience was comprised of females from the Millennial Generation. Additionally, the 2012 rendition of the music festival softened the negative trend in US arrivals, as well as drew repeat festival attendees thus increasing potential tourist loyalty to the island.

Empirical literature related to the Aruba Soul Beach Music Festival: research findings

There are only a handful of empirical studies that have examined the topics that emerged in the aforementioned results from the Aruba Soul Beach Music Festival of 2011 and 2012. Croes and Lee (2015) investigated whether gender played a role in attendees' motivations to attend a music festival in a SID. The results of their study revealed that gender does indeed contribute to an individual's travel choice and leisure consumption patterns at a destination. Thus, gender differences (particularly women) are a relevant topic for festival and destination managers to examine when designing attractions and tourist product offerings. The context of the Croes and Lee (2015) study specifically investigated women that opted to attend a SID music festival.

In relation to music festivals attracting younger tourists to the island of Aruba, Rivera *et al.* (2015) examined the use of SID music festivals within the experience economy framework to determine if music festivals could be used as experiential attractions to attract a younger cohort replacement audience to that of senior tourists. The Rivera *et al.* (2015) study also investigated the experiential domains that are necessary to positively influence the overall experience of a younger generation. Their study found that if the traditional experience economy framework is adjusted to include economic value as well as education, entertainment, escapism and aesthetics then it is possible for a music festival to draw a younger tourist base to an island destination. Semrad and Rivera (2016) confirmed these findings in the context of Generation Y and SIDs.

Curacao North Sea Jazz Festival (years 2010–2015)

2010 Curacao North Sea Jazz Festival: Uncovering the Potential of a New More Affluent and Loyal Market

In the report entitled *Uncovering the Potential of a New More Affluent and Loyal Market* by Croes *et al.* (2010), the investigators sought to determine the value of the music festival to the attendees and to determine its economic impact to the island economy. Overall, the festival organizers spent $3,000,000 to organize the event and $700,000 was directly spent on the island. The majority of the money that was spent to organize the event was used to pay top artists to perform in the first-time production of the festival.

The overwhelming majority of the festival attendees felt that their experience at the event exceeded their expectations and thought that the music festival represented value for the money they spent at the festival. The majority of the attendees (86%) indicated that their primary reason for visiting Curacao was to attend the festival and (70%) indicated that they were planning to attend the festival next year. The types of tourists that the music festival attracted to Curacao were not traditional tourists that frequented the island. In other words, similar to the Aruba Soul Beach Music Festival, the Curacao North Sea Jazz Festival attracted more affluent, higher educated, first-time tourists that spent more money while on the island than traditional tourists. Thus, from a destination perspective, the island procured significant economic benefits from the festival attendees' spending while on the island.

The total economic benefits to the island amounted to $4,000,000 in 2010. The tourists that attended the festival spent an average of $119 per day per person and stayed four days on the island. The multiplier effect was 1.91. This market segment performed better than the traditional Curacao tourists who spend $107 per day and generate a multiplier effect of 1.83.

2011 The Signature Event in Curacao: A Source of Brand Equity and Economic Significance

In the report entitled *The Signature Event in Curacao: A Source of Brand Equity and Economic Significance* by Rivera and Croes (2011), a similar study was conducted to determine the attendees' perceptions of the music festival, as well as the economic impact of the festival to the island economy. The organizer increased the spending to plan the event from $3,000,000 to $4,900,000. The total economic benefits to Curacao from the festival increased 311% for a total amount of $16,338,601 from the previous year's production.

The festival attracted more than twice the amount of tourists than the previous year. However, this year's production led to tourists staying on the island for 7.3 days and spending $233.00 per day (more than half the amount spent in 2010). Again, the types of tourists arriving to the island deviated from Curacao's traditional tourists in terms of their annual salary, their level of education, and the amount of money they spent per day. They also deviated in that their preferred choice of accommodation were hotels as opposed to apartments or vacation rentals. Thus, the festival boosted the occupancy rates of the local hotels and the overall consumption that took place in restaurants.

Since many of the tourist festival attendees were repeat attendees from the previous year, a new finding emerged regarding the value of the festival. Many of the festival attendees indicated that they felt the festival allowed them to immerse themselves within Curacao's culture, to make friends and share time with people from different backgrounds. They also indicated that the price structure of the festival was appropriate, meaning they received value for the money they spent. Thus, a majority of the 2012 festival attendees indicated that they would recommend both the festival and the destination to friends and family.

2012 Curacao North Sea Jazz Festival: Exceeding Expectations & Evoking Synergies for Future Growth

In the report entitled *Exceeding Expectations & Evoking Synergies for Future Growth* by Rivera and Croes (2012), the investigation sought to follow a similar line to that of the previous years in terms of assessing festival attendees' satisfaction with the event and the economic impact to the island. The 2012 investigation also explored specific market challenges that may hinder Curacao's ability to continue to grow the festival in future years' productions.

The festival organizer invested $5.6 million into the planning of the 2012 festival. This was a 14% increase from 2011. In 2012, the festival churned out nearly $20 million to the local economy, which was nearly a 20% increase over the previous year. The festival attendees spent an average of $299 per day per person. This spending behaviour increased over the first two renditions of the festival thus indicating the close correlation between the levels of satisfaction and spending behaviour. The sector that benefited the most from the increase in spending was the hotel industry. Hotels saw an increase in spending from $184 a day to $515 per day – which is a 180% increase. The growing popularity of the festival demonstrated a change in ticket purchasing behaviour where average festival attendees purchased their tickets three or more months in advance (compared to less than three months in advance in 2010).

The growth of the music festival also introduced more new market segments to the island of Curacao. A new Hispanic market segment emerged as showing increased interest in attending the festival. Also, 'island hoppers', (i.e. tourists in the Caribbean that learn of the festival and 'hop' islands to attend the festival) appeared at the 2012 festival production. The new market segments that emerged in the 2012 study were mostly first-time festival attendees, as well as first-time visitors to the island of Curacao. Thus, the Curacao North Sea Jazz Festival continued to penetrate new markets, acquire new tourist arrivals and provide Curacao with the opportunity to increase repeat visitation.

Some supply constraints were revealed in the 2012 study. These supply constraints present some challenges to Curacao. While the success of the festival is inducing greater tourist loyalty, the supply constraints (e.g. airlift and hotel rooms) may limit the festival's continued growth. Additionally, as the quality level of the festival continues to increase a growing gap between the quality of Curacao's tourism product and the festival becomes more evident. This means that festival attendees may experience a 'wow factor' from the festival but may become complacent with the overall quality of Curacao's tourism product. Thus, the 2012 report revealed that Curacao destination managers might need to work to build the island's image, market position and quality of tourism offerings. It is also necessary for destination managers to concentrate on developing strong partnerships within the tourism industry to enhance the overall quality of the product.

2013 Curacao North Sea Jazz Festival: Through 'The Looking Glass'

The 2013 report entitled *The Curacao North Sea Jazz Festival: Through 'The Looking Glass'* by Rivera and Croes (2013), opted to specifically concentrate on the spending behaviour and impact of tourists attending the festival. For the first time, the festival observed a dip in its overall economic contribution to the island's economy. This is likely due to a dip in attendance by 2000 festival attendees that may have occurred because a major airline went bankrupt. Of the 4332 tourists that attended the festival, it was reported that the average time spent on the island was 5.49 days (a decrease of two days).

This decrease in the length of time spent on the island may be due to the majority of the attendees being repeat attendees and tourists to Curacao, as well as hotels inflating the price of rooms due to the high demand in previous years from the festival (aggressive pricing). Also, if one considers that first-time visitors generally tend to spend more time at a destination and that the festival has more repeat attendees than first-time attendees, then the reduced length of time spent in Curacao could be explained.

A high note from the 2013 festival production is that the tourists that attended the festival spent more money per day per person than the previous years. In 2012, the festival attendees spent $299 per day per person. In 2013 this figure increased to $328 per day per person (a 10% increase). The continued increase in spending per day is testimony to the affluent market segments that the Curacao North Sea Jazz Festival has been able to attract over time.

An interesting finding in the 2013 study was that only 17% of the tourists' spending that attended the music festival was absorbed by the festival organizers. This means that 83% of tourists' spending went to suppliers and service providers in Curacao. Thus, although 2013 was not the festival's best production year it still made a significant economic contribution to the island.

The 2013 findings confirm what was anticipated in the 2012 study, which is that the growing gap between the quality of the festival performance and the overall tourism product in Curacao hampered the ability of the festival to continue its growth. Thus, while the festival benefits Curacao's image the island does not benefit the festival in the same fashion. In addition, the results from the 2013 study seem to indicate that the festival may be reaching a point of market saturation.

2014 The Curacao North Sea Jazz Festival: A Lustrum Churning Economic and Promotional Opportunities for Curacao

The 2014 report entitled *A Lustrum Churning Economic and Promotional Opportunities for Curcao* revealed that since its inception the festival has generated $75 million to Curacao's economy. The report also demonstrates that had the festival not taken place in the month of August (2014) that the hotel industry would have lost $7 million of its revenues.

Beyond the economic benefits that the festival has brought to Curacao, the event has also contributed to the welfare of the island by enhancing the image of the destination. For example, 80% of the 2014 festival attendees reported that their image of the island improved after attending the event. The festival also generated widespread media exposure, which has indirectly generated millions in promotion for Curacao in international markets.

In 2014, it was clear that the festival acquired a distinct brand of its own, apart from the island, when 86% of the attendees indicated that they travelled to Curacao because they wanted to attend the event; and the majority of the attendees had been to Curacao in the past for purposes of the festival. The economic contribution of the 2014 festival amounted to just over $20 million – an improvement from the 2013 rendition.

While the festival attendees clearly indicated in the 2014 data that they were highly satisfied with the festival experience, there was a misalignment between the quality of the festival and the destination's tourism product. In other words, the value of the tourism offerings at the destination does not match the value or quality level of the festival. This is a dangerous realization in that without the production of the festival it is likely that these tourists would not come to Curacao. Thus, it seems that in the 2014 festival production that a brand equity gap may be occurring between the music festival and the host destination. This information may present a looming marketing concern for destination managers especially when considering that the festival is used as an opportunity for Curacao to gain market exposure and entry with particular market segments.

2015 Curacao North Sea Jazz Festival and Destination Convergence

The 2015 festival report entitled *Curacao North Sea Jazz Festival and Destination Convergence* by Croes *et al.*, (2015), revealed that 7,216 tourists attended the event. An interesting finding emerged this year in that nearly one out of four attendees were single females (a similar attendance trend to the Aruba Soul Beach Music Festival). The majority of the attendees indicated that their main reason for travel to Curacao was to attend the festival. Attendees' festival satisfaction (95% satisfied) and intention to recommend the festival remained strong like all previous years.

The spending per person per day was $311 where hotels and restaurants captured nearly half of the total attendees' spending. The other categories that benefitted were car rentals and nightlife. The economic contribution to the island surpassed all other years, and for the first time, the 2015 study began to show a convergence between the destination's tourism offerings and the quality of the festival. Evidence of this is found in the attendees indicating that had the festival not occurred they would still be likely to travel to Curacao.

Empirical literature related to the Curacao North Sea Jazz Festival: research findings

The Curacao North Sea Jazz Festival heavily concentrated on the economic benefits to the island's larger economy that could be derived from the festival. As previously mentioned, Kim *et al.* (2010) referenced that one of the major research streams pertaining to festivals as tourism attractions is the economic contribution that may be realized by the destination that hosts the event. In the context of music festivals and SIDs, this economic contribution is derived via the destination diversifying its experiential offerings resulting in an increase in tourist arrivals that is reflected in increased foreign exchange and economic activity that is incurred from forward and backward economic linkages.

A relatively new application of a research framework that has emerged to explain how music festivals may benefit SIDs' economies is that of the OLI paradigm. Ferdinand and Williams (2013) discuss the potential relevance of the OLI paradigm in the investigation of internationalization benefits of festivals in general. Rivera *et al.* (2015) operationalized and tested the framework in the context of music festivals occurring within SIDs.

The Rivera *et al.* (2016) study investigated the Curacao North Sea Jazz Festival during the years of 2010–2015 by way of a macro economic analysis. The analysis enabled the researchers to determine if a SID could incur internationalization benefits by way of a music festival. These benefits would include both inward and outward internationalization effects. Inward internationalization includes the business transactions that take place on the island by the tourists that are attending the festival (e.g. hotel payments) whereas outward internationalization includes marketing activities that occur in foreign markets, which drives tourist arrivals to the island for purposes of consuming the festival. The study found that music festivals do make a significant and positive economic impact and do provide internationalization benefits for a small economy. These benefits are achieved when SID festival organizers produce a high-quality music festival in a destination where the tourism offerings match the quality of the festival.

It is important to recall that in a confined economy there are two audiences that can attend a music festival: those that reside on the island and those that must arrive by travel (tourists). Thus, the arrival of tourists who seek to attend a high-quality music festival will increase tourist receipts, thus increasing the largest export category (i.e. foreign exchange earnings) for a SID economy. The OLI paradigm provides an empirical framework to explain the continued economic success and impact to Curacao's economy.

11.5 Conclusion

The current chapter sought to explore the positive contributions that music festivals could make to a SID in terms of its tourism industry. The chapter reviewed longitudinal research findings from eight different music festival productions that occurred in Aruba and Curacao (i.e. Aruba Soul Beach Music Festival and the Curacao North Sea Jazz Festival). As far as it is known, there is no other comprehensive reporting on longitudinal SID music festival research findings available.

The longitudinal research findings provided within this chapter reveal that the production of high quality SID music festivals could provide destination managers with the opportunity to attract and acquire new tourist market segments with the possibility of spending more money during their island stay. Thus, a stimulation of positive economic impact through increased tourist receipts and foreign exchange could occur. Given that the Caribbean region is emerging as a popular market place for music festival productions, future research should strive to test the five primary research streams that emerged when destinations positioned festivals as tourism attractions. This testing should occur within the context of SIDs and specifically music festivals. Moreover, future research should investigate the following: (i) attendees' motivations for consuming SID festivals; (ii) residents' and/or attendees' perceptions of SID festivals; (iii) economic impacts of SID festivals; (iv) methods used to assess economic impacts of SID festivals; and (v) characteristics that influence consumption patterns of SID festival attendees.

References

Andersson, T.D. and Getz, D. (2009) Tourism as a mixed industry: Differences between private, public and not-for-profit festivals. *Tourism Management* 30(6), 847–856.

Croes, R. (2011) Measuring and explaining competitiveness in the context of small island destinations. *Journal of Travel Research* 50(4), 431–442.

Croes, R. and Lee, S. (2015) Women at a music festival: Biological sex defining motivation and behavioral intentions. *Event Management* 19(2), 275–289.

Croes, R. and Rivera, M.A. (2012) The Aruba Soul Beach Music Festival: Getting the Fundamentals Right.

Croes, R.R., Rivera, M.A. and Semrad, K. (2010) Uncovering the Potential of a New More Affluent and Loyal Market: The Curacao North Sea Jazz Festival. *Dick Pope Sr. Institute Publications* 14. Available at: http://stars.library.ucf.edu/dickpope-pubs/14 (accessed 9 October 2017).

Croes, R.R., Rivera, M.A. and Olson, E. (2011) *The Economic and Tourism Potentials of the Aruba Soul Beach Music Festival.* Aruba Tourism Authority, Aruba.

Croes, R.R., Rivera, M.A. and Semrad, K. (2015) Curacao North Sea Jazz and Destination Convergence: A Harbinger Beckoning? *Dick Pope Sr. Institute Publications* 39. Available at: http://stars.library.ucf.edu/dickpope-pubs/39 (accessed 9 October 2017).

Easterly, W. and Kraay, A. (1999) Small states, small problems? Policy research working paper No. 2139. Development Research Group, World bank, Washington, DC.

Felsenstein, D. and Fleisher, A. (2003) Role of public assistance and visitor expenditure. *Journal of Travel Research* 41(4), 385–392.

Ferdinand, N. and Williams, N. (2013) International festivals as experience production systems. *Tourism Management* 34(1), 202–210.

Frey, B.S. (1994) The economics of music festivals. *Journal of Cultural Economics* 18(1), 29–39.

Kim, S., Prideaux, B. and Chon, K. (2010) A comparison of results of three statistical methods to understand the determinants of festival participants' expenditures. *International Journal of Hospitality Management* 29(1), 297–307.

Morgan, M. (2008) What makes a good festival? Understanding the event experience. *Event Management* 12(2), 81–93.

Prentice, R. and Anderson, V. (2003) Festival as a creative destination. *Annals of Tourism Research* 30(1), 7–30.

Richards, G. and Wilson, J. (2005) Developing creativity in tourist experiences: A solution to the serial reproduction of culture? *Tourism Management* 27(6), 1209–1223.

Rivera, M.A. and Croes, R.R. (2011) The Signature Event in Curacao: A Source of Brand Equity and Economic Significance. *Dick Pope Sr. Institute Publications* 12. Available at: http://stars.library.ucf.edu/dickpope-pubs/12 (accessed 9 October 2017).

Rivera, M.A. and Croes, R.R. (2012) CNSJF: Exceeding Expectations and Evoking Synergies for Future Growth. *Dick Pope Sr. Institute Publications* 1. Available at: http://stars.library.ucf.edu/dickpope-pubs/1 (accessed 9 October 2017).

Rivera, M.A. and Croes, R.R. (2013) Curacao North Sea Jazz Festival: Through "The Looking Glass". *Dick Pope Sr. Institute Publications* 18. Available at: http://stars.library.ucf.edu/dickpope-pubs/18 (accessed 9 October 2017).

Rivera, M.A., Croes, R.R. and Semrad, K. (2014) The Curacao North Sea Jazz: A lustrum churning economic and promotional opportunities for Curacao. *Dick Pope Sr. Institute Publications* 37. Available at: http://stars.library.ucf.edu/dickpope-pubs/37 (accessed 9 October 2017).

Rivera, M., Semrad, K. and Croes, R. (2015) The five E's in festival experience in the context of Gen Y: Evidence from a small island destination. *Revista Espanola de Investigacion de Marketing ESIC* 19(1), 95–106.

Rivera, M., Semrad, K. and Croes, R. (2016) The internationalization benefits of a music festival: The case of the Curacao North Sea Jazz Festival. *Tourism Economics* 22(5), 1087–1103.

Schubert, S., Brida, J. and Risso, W. (2011) The impacts of international tourism demand on economic growth of small economies dependent on tourism. *Tourism Management* 32(1), 377–385.

Semrad, K. and Bartels, J. (2014) An inward look using backward economic linkages in a developing country: The case of Puntarenas, Costa Rica. *Worldwide Hospitality and Tourism Themes* 6(3), 244–260.

Semrad, K. and Rivera, M. (2015) A destination performance analysis through the comparison of tourists and stakeholders' perceptions: The case of Curacao. *Journal of Tourism Hospitality* 4(171), 2–6.

Semrad, K. and Rivera, M. (2016) Advancing the 5E's in festival experience for the Gen Y framework in the context of eWOM. *Journal of Destination Marketing & Management*. Available at: http://dx.doi.org/10.1016/j.jdmm.2016.08.003 (accessed 9 October 2017).

Srinivasan, T. (1986). The costs and benefits of being a small, remote, island, landlocked, or mini state economy. *The World Bank Observer* 1(2), 205–218.

12 Developing a Network Analysis Methodology for Island Tourism Research

Michelle McLeod*

The University of the West Indies, Mona, Jamaica

12.1 Introduction

Island environments are ideal microcosms to research the tourism phenomenon as island settings provide an opportunity to develop the research process, and to explore and examine what is yet to be known about tourism. In that regard, methodological approaches to study tourism and how network methodologies in particular can improve understanding of island tourism specifically and improve understanding about tourism in general have been considered.

With island environments having dominant tourism industries it is important to understand how to improve the performance of tourism since there are island states that are not benefiting from tourism. McLeod, Lewis and Spencer (2017) note that although the Caribbean has a dominant tourism industry across several islands, the benefits being derived, particularly by locals, have been limited and they propose a research agenda including re-developing a Caribbean tourism identity, access and profitability, to improve the fortunes of Caribbean tourism. Naturally, with these challenges research methodologies need to be designed that are aligned to the resolution of these issues in island environments.

A focus on developing island tourism can also provide benefits for residents. Tourism research studies have demonstrated that islands' residents have benefitted from tourism development. For instance, community integration in tourism led to socio-economic benefits for residents (Mitchell and Reid, 2001), island residents recognize the positive impacts of tourism although there are some negative influence of tourism (Nunkoo and Ramkissoon, 2010) and island residents appreciate tourism for its economic gain and provision of quality experiences (Brown, Cave, Moyle, Croy and Weiler, 2010).

Challenges of tourism development within islands have been addressed by application of the concept of sustainability. The concept of sustainable tourism has gained

*E-mail: michelle.mcleod@uwimona.edu.jm

momentum in the tourism literature. The establishment of the *Journal of Sustainable Tourism*, and the 2017 International Year of Sustainable Tourism for Development declared by the United Nations General Assembly point to the growth about this topic. Despite this, the process of achieving an optimal state of sustainable tourism is still unknown and this speaks to a gap in methodological practices in the study of tourism. Sustainable tourism measurement has progressed conceptually and the development of tools such as the Barometer of Tourism Sustainability (BTS) has been adopted (see Ko, 2005; Senna *et al.*, 2015). Nonetheless, in their 25 years article to review the *Journal of Sustainable Tourism*, methodological advancement stops short, as authors suggest that research methods are narrow and often single-destination case studies (Bramwell, Higham, Lane and Miller, 2017). New methodologies can assist with understanding of the tourism phenomenon, its sustainability and in particular how to improve the fortunes of island tourism.

Issues that result in the occurrence of the state of affairs in the tourism industry are interconnected. For instance, although views about the effects of climate change on islands are well documented (Belle and Bramwell, 2005), but without building methodologies that can predict the likely occurrences of these effects on the tourism industry, the sustainability of the industry is challenged. Tourism is a dynamic sector and studying one set of variables at a time brings limits to the advancement of knowledge. By the time all possible variables have been examined and explored and likely occurrences have been determined, one unknown that has now been revealed changes knowledge. The dynamic nature of tourism and the interconnectedness of tourism processes suggest a need to advance the creation of knowledge in tourism studies at a faster rate.

New approaches to tourism research are needed to advance knowledge on the principles and practices of tourism that will provide increased benefits for tourism in warm-water island destinations, which have year-round tourism industries. These new approaches should consider the usefulness of multidisciplinarity as multidisciplinarity may be slowing down the creation of new knowledge in tourism, since basing tourism knowledge on multiple disciplines takes tourism theory in several directions and reduces the chance for the establishment of foundational tourism theory that can support the provision of real-world solutions to fledgling tourism industries in island environments.

The chapter therefore reviews tourism paradigms, methodologies and research approaches within existing research paradigms. The particular characteristics within island environments that would guide the research process have been explored. Also, an exploration is conducted of possible approaches for studying tourism systems such as an island tourism destination and methodological approaches suggested to conduct island tourism research with the argument that an advancement can be made in optimizing sustainable tourism through utilizing system methodologies.

12.2 Tourism Paradigms, Methodologies and Research Approaches

Tourism by nature is not a complex term and therefore should not be that complicated to study but there seems to be a lack of focus. McLeod (2005, p. 2) notes, 'The root of the word tourism is related to the word tour ... tourism is the science of touring ... the study of tourism involves an understanding of the motives, practice and benefit of the tour.' Yet tourism paradigmatic discourses are perhaps avoided because of the 'indiscipline of tourism' (Tribe, 1997), in that the study of the tourism phenomenon utilizes sporadic methodologies from hither and thither with no particular direction to result in a grand theory of tourism. Tribe (1997) rejects the idea that tourism is a discipline and proposes two fields, the business of tourism and the non-business aspects of tourism. He organizes tourism studies as a field involving four epistemologies: multidisciplinarity, general interdisciplinarity, business interdisciplinarity and extradisciplinarity. While tourism studies focus on the management of tourism with strong emphasis on marketing management, Benckendorff and Zehrer (2013) based on their network analysis of tourism research, note that the 'business of tourism', except for consumer behaviour, is not a substantial part of the tourism literature.

Dann (1997) in addressing paradigms in tourism research notes that tourism is a multidisciplinary field. Almost 20 years later, Tribe, Dann and Tazim (2015) writing about paradigms in tourism research, a trialogue, note that the proposition that tourism is a field of study and not a discipline has not been falsified. The authors suggest that the field of tourism began from certain disciplinary orientations (perhaps economics and geography) and then progressed using new methodological paradigms such as constructivism, phenomenology and critical theory. Nonetheless, one contribution of paradigms is the formulation of conceptual frameworks (Tribe, Dann and Tazim, 2015). This last point suggests that tourism studies should be grounded in a paradigm but the view that tourism research draws mostly from a qualitative approach (Tribe, Dann and Tazim, 2015) is debateable.

Echtner and Jamal (1997) suggest that tourism theory is fragmented and weak based on research approaches crossing many disciplines. For tourism studies to have disciplinary status, a group of researchers would need to break away from disciplinary conventions utilized in tourism and develop a 'distinct disciplinary matrix' for tourism studies in the way of a Kuhnian 'scientific revolution' (Echtner and Jamal, 1997, p. 876). The dialogue about whether tourism studies has disciplinary basis should take on a different focus, one that involves a focus on the methodology, which is the way of knowing, and new paradigms developed, if non-existent, to justify research approaches.

Methodologies are built during a process of enquiry that is based on a particular belief system of knowing about phenomena. Belief systems are paradigms about knowing, and Guba (1990) makes the distinction between ontology, epistemology and methodology as an ontology is the nature of the 'knowable', epistemology is the relationship between the knower and the 'knowable', and methodology is the way to discover the 'knowable'. Ways of knowing about the tourism phenomenon have been largely treated as a social science with multi disciplinary approaches being adapted as new techniques and ideas emerge.

Tourism studies have evolved using various research frameworks. A research framework is a methodology. Methodology is a systematic construction to systematically create new knowledge within the research process as a scientific endeavour. In principle, a research study is designed based on axiological and ontological standpoints to achieve the aim of the research study and to allow for the answering of the research question or questions. It is the research study design that is the methodology, and this has an epistemology as to the role played by the researcher in knowing. The research design sets out first the research location, and then data collection and analytical techniques. Generally, these activities occur within three main research approaches, i.e. quantitative, qualitative and mixed methods research designs. Ongoing discourses, as to whether tourism studies follow or lend itself more to one design or another, and whether the tourism discipline should be developed within one particular context or another are needed.

Context specificity seems to be the basis for qualitative research approaches to derive the rich data within a range of circumstances to better understand the nature of the subject under study. Quantitative research approaches seek to generalize findings that are applicable to a range of contexts. Mixed methods seem to utilize the advantages of both approaches to build a research design that utilizes the strengths of one approach to avert the weaknesses of another approach.

Slevitch (2011) notes fundamental differences between qualitative and quantitative methodologies and advocates for a re-focusing on the ontological and epistemological assumptions of both approaches rather than the differences in the methods of each approach. Slevitch (2011) notes that quantitative research relates to positivism that is based on a realist assumption that a 'truth' exists to be revealed using objective, value-free methods to generalize findings. Qualitative research is idealistic, based on interpretivist and constructivist perspectives, multiple realities are explored using subjective, value-based methods to contextualize findings (Slevitch, 2011). Based on this distinction, Slevitch (2011) suggests adherence to a particular paradigm and the non-integration of methodologies although methods can be integrated.

Davies (2003) explores methodologies in tourism to determine a framework that combines both quantitative and qualitative research as there are shortcomings with methods to understand the tourism business environment. An integration between the two research approaches is needed particularly as quantitative approaches based on positivism suggests that 'structures do not change or evolve' (Davies, 2003, p. 100). Tourism is a dynamic phenomenon that involves a range of interacting relationships, links and connections across time and space and hence the reason that positivism constrains understanding of the phenomenon.

Simply combining quantitative and qualitative approaches as methodological triangulation has its drawbacks in that the two approaches are founded on two different paradigms (see Slevitch, 2011). However, Davies (2003) suggests a framework that involves complementarity of quantitative and qualitative data to advance understanding of tourism businesses, and cites Miller and Crabtree's (1994) work about four research designs that can be adopted in tourism research including concurrent, sequential, nested and combination research designs. Also, Iaquinto (2016) showed that application of mixed methods by combining participant observation, interview and questionnaires to reveal sustainable practices of backpackers provided different results but was necessary to obtain a thorough answer to the research question.

12.3 Rationale for a New Methodological Approach in Island Tourism Research

With island tourism being a special form of tourism, particular attention should be given to the conduct of the research process for effective results that can improve the performance of island tourism. Island tourism studies are challenged with small samples, various cultural contexts, resource constraints, and a commonality is that these tropical islands are largely developing countries. Research is a tool that identifies the issues within tourism and creates an opportunity for resolution; however, research activities require resources. Island environments tend to be smaller in size, but Pearce (2008) notes that islands have limited resources for physical development and therefore planning is an important activity to minimize the adverse impacts of tourism development. For the research process a distinction has to be made between warm- and cold-water island tourism destinations. Baldacchino (2006) suggests that the nature of the environment has resulted in a limited number of tourists visiting cold-water island destinations whereas warm-water island destinations attract the masses.

Islands have unique characteristics that warrant methodological advances for improving tourism performance. Sharpley (2012, p. 167) notes that islands are associated with 'remoteness, separateness, difference and the exotic, they are the stuff of romance and adventure, of fantasy and escape, of "otherness"'. Several issues have been noted about island tourism, including dependence, economic and political weakness, environment distinctiveness, isolation, monoculture, peripherality, risks, smallness and vulnerability (Harrison, 2001; Coccossis, 2002; Lim and Cooper, 2009). Conversely, Scheyvens and Momsen (2008); these are all concerned about the characterization of small island states and suggest that discourses of fragility and vulnerability suggest a need for external intervention in the affairs of island sates. They note the strengths of these nations including 'small is beautiful', 'small islands are good economic performers', 'high levels of cultural, social and natural capital', 'respect for traditional, holistic approaches to development', 'strong international linkages', and 'political strength' (Scheyvens and Momsen, 2008). Whether there are positive or negative views of island characteristics the research challenge would be the effect of these characteristics on the findings of either a quantitative or qualitative research study.

With natural tending characteristics that favour tourism development, it is difficult to understand why several island tourism destinations have failing industries. For instance, some issues facing Caribbean tourism development include a mono-product area, a tourism system that has not adequately developed towards local community development and the mass tourism arrivals that threaten the environment (Zappino, 2005). In addition, Sharpley (2014) explores governance of tourism in small island developing states (SIDS) and points to the important role of government support for tourism development. Using the case of Zanzibar, Sharpley (2014) notes particular challenges for tourism development, including a lack of investment, planning and policy failures, institutional confusion and political interference/patronage. While islands are small land masses these seem to be pervaded with a range of issues relating to tourism development that have not been holistically addressed by tourism researchers.

Some tourism researchers have suggested several solutions. Sheldon (2005) proposes the following approaches to overcome the challenges of island tourism, including long-term stakeholder-involved planning, empowerment of the island community and culture, environmental management, visitor management, knowledge and information systems, accessibility and transportation, and marketing and market diversification. Charlie (2014), referring to small island developing destinations (SIDD), draws on the context of small Indonesian island tourism destinations to propose a framework for an environmental governance network to manage the process of small island tourism through involvement, engagement and collaboration. Lim and Cooper (2009) suggest a new life cycle model to optimize island tourism development. Optimization involves identification and development of indicators and triggers of island tourism development. Zhang, Lui and Wan (2008) considered the functional zoning of island tourism and categorize regions from preferred development to prohibited exploitation. According to researchers, tourism development is influenced by the island resources, carrying capacity, exploitation density and development potential (Zhang *et al.*, 2008).

While solutions are important, the effects of several interventions at the same time are unknown and therefore another approach is needed. Certainly, islands are smaller land spaces that can provide a comprehensive illustration of the evolution of tourism dynamics within time and space. Yang, Ge, Ge, Xi and Li (2016) explored such evolutionary dynamics in an island context by measuring changes of the island tourism landscape. While this physical mapping can assist with tourism planning activities, it is also important to understand how such an evolution contributed to patterns of tourist flows and tourist flows occur within a system.

The systemic nature of tourism relates to the flows of tourist activities during touring. A flow is a movement from one point to another and this causes an interaction between the tourist and different points along the tour product. The tourism system therefore starts at a place, flows occur, interconnections happen and this process can be mapped. A tourism system for an island can be mapped to determine the interconnections between the tourist and the tour; this is the core basis for the nature of tourism phenomenon and therefore should form the essence of a methodological advancement in tourism research. A breakdown of the structure of a tourism system provides an opportunity for the study of tourism systems in a relatively small bounded area of an island and also provides a basis for island tourism growth and development.

12.4 Possible Pathways for Designing Tourism System Methodologies

A research study of the tourism phenomenon lends itself to the adoption of methodologies that involves systems theory. A tourism system interconnects the elements of a tourism experience and hence the reason system methodologies have been popularized to understand the tourism phenomenon. Leiper's (1979) tourism system utilizes systems thinking to develop a new definition of tourism that refers to tourists, generating regions, transit routes, destination regions and a tourist industry comprises several actors with functional and spatial connections across the system (Leiper, 1979). Carlsen (1999), drawing from Leiper's (1979) work, calls for a systems approach to island tourism destination management because of the shortcomings of the reductionist approach to monitor visitor flows or the measurement of relationships between variables such as expenditure and employment. He suggests a soft systems methodology (SSM) that is a practical approach to understand macro issues such as sociocultural, political and environmental aspects of an island that can affect tourism with the view of proposing solutions to real-world problems (Carlsen, 1999).

Interconnections within the tourism systems are networked. In that regard, several tourism researchers have begun to detail the workings of networks in tourism. Merinero-Rodríguez and Pulido-Fernández (2016) advocate for the tourism networks line of research as it has facilitated understanding and revealed new characteristics of tourism relationships. According to Merinero-Rodríguez and Pulido-Fernández (2016, p. 129), network analysis allows the consideration of four aspects of the tourism phenomenon:

> i) the factors and conditions that make it possible for interactions between the different stakeholders involved in tourism activity to occur; ii) the structure of the relationships between said components, or their mathematical properties; iii) the productive bonds of tourism in the destination; and iv) the effects and impacts of the characteristics of relationships on tourism activity and, in accordance with those effects, the assessment and proposal of actions that will affect these relationships with a view to improving the functioning of tourism activity.

Tourism network characteristics in business and personal networks have been examined by McLeod, Vaughan and Edwards (2010) to reveal the importance of network structure in relation to network density and the positioning of network actors in relation to centrality and the knowledge potential of tourism actors. Thereby network analysis has its own paradigm in the sense that both variables and their influences are studied at the same time to understand systemic features of a network pattern. Pavlovich (2003) suggests that network pattern can provide a causal explanation of the evolution and development of a tourism destination network. Based on this, a network analytical approach to tourism studies provides a real opportunity to provide additional insights into the tourism phenomenon.

Conceptually, network analysis can be utilized but its practical application to tourism research involves a number of authors studying the nature of tourism systems at the destination level using social network analysis (SNA) (Scott, Cooper and Baggio, 2008; Baggio, 2010; Baggio, Scott and Cooper, 2010b; McLeod, Vaughan and Edwards, 2010); however, few focus on destination flows in relation to the tour product (Baggio and Sainaghi, 2016; Shih, 2006). Baggio and Sainaghi (2016) focus on the nature of tourism demand within a tourism destination using network analysis and revealed the dynamic nature of tourism flows. Shih (2006) compared the network characteristics of drive tourism destinations and notes that SNA is an appropriate technique for destination analysis as it provides measures regarding the structural patterns of connected systems.

SNA complements understanding of tourism systems. SNA is a theory and methodology of revealing relational structures. Borgatti and Halgin (2011) note that SNA is about the antecedents and consequences of structural patterns. Within SNA the nodes are distinct entities that are related to each other through interactions or flows called ties. These relationships may be strong or weak and Granovetter (1973) theorizes that there is strength in a weak tie as such a tie interconnects or bridges the network by reaching nodes that are otherwise unconnected to a focal node. The network forms a pattern and such a reality has been theorized by Giddens (1984) to be structuration. Giddens (1984) structuration theory suggests that there is duality of influence as the nodes form a structural pattern and that structural pattern also affects the nodes. Network influence can be analysed based on its shape. Based on studies conducted by Bavelas and Leavitt in the 1950s, the shape of the network is an indicator of its consequences, as a circle-shaped network that is decentralized is the worst, Y-shaped and chain-shared structures limit centralization, and a star-shaped network is the best, for speed and accuracy of solutions (Borgatti *et al.*, 2013). SNA theory and methodology is an appropriate paradigm within which to study tourism systems. Tourism systems involve a pattern as the touring process occurs through the interactions of largely airlines, hotels and attractions that operate within scheduled times and that tourists engage with as part of their tour.

In essence, tourism systems involve interactions between human and non-human actors. Based on this, actor-network theory (ANT) (Latour, 1996, 2005; Law, 2007) has been adopted by several authors to understand the nature of tourism (Ren, Jóhannesson and Van der Duim, 2012). ANT provides a theoretical lens to explore both human and non-human elements of a tourism system at the same time. ANT has a role in understanding the tourism system as application of ANT reveals the important role of non-human agency to influence consequences of a network of actors. Authors have utilized ANT to explore relationships in tourism. McLeod and McNaughton (2016) successfully mapped an emergent open data ecosystem for Caribbean tourism data utilizing ANT as theoretical foundation and SNA as theory and methodology. Hummel and van der Duim (2016) applied ANT to explore modes of ordering within an organization managing tourism projects. Padget, Dimanche and Mounet (2010) utilized ANT to explore innovation in a tourism company using a multi-methods qualitative approach and proposed that a winter sports activity involves the combination of both humans such as hosts and activity providers, and non-humans such as ski lifts and skis, and this combination changes depending on the type of customer. Work using ANT to understand the tourism phenomenon has been advancing; however, there is a gap in measuring the entire tourism system. ANT has not been applied to explore a tourism system and SNA to measure such a system to understand the nature of tourism and to reveal the antecedents and consequences of such a system of actors.

12.5 Network Analysis Methodology in Tourism Management Research

Network analysis for tourism research has utilized both qualitative and quantitative approaches (Scott, Baggio and Cooper, 2008) but reliance on qualitative data has been popularized in tourism management research. Manderscheild (2016) notes a similar trend to rely on qualitative data in mobilities research and argues for quantification to handle the shortcomings of qualitative data. Measurement of the quantitative data within the tourism destination system is important for understanding how the system functions. Bonetti, Petrillo and Simoni (2006) suggest that tourism destination systems are linked based on the degree of interdependence among the stakeholders as determined by the density of relationships and the degree of centralization of the system governance functions. Pechlaner, Volgger and Herntrei (2012) used a qualitative approach focussed on the interdependencies between DMO performance and destination performance to reveal the level of cooperation within the destination. Network science allows for the modelling of management actions on the tourism system to refine strategies and governance frameworks (Baggio, Scott and Cooper, 2010a). Baggio, Scott and Cooper (2010a, p. 57) propose five governance components that can be based on network measures:

> i) power and salience based on centrality measures; ii) cooperation based on the clustering coefficient modularity; iii) knowledge management based on local and global efficiency; iv) prediction based on simulations of addition and deletion of network elements and evaluation of the effects on dynamic processes, and v) scenario analysis based on simulation of dynamic processes using network connections as a substrate and scenario building by modifying network characteristics or process input parameters.

Thus, advances can be made in tourism research by utilizing a network perspective as such an approach facilitates understanding of the operation of a tourism system, its structures and patterns and thereby realistic plans and predictions can be made to improve the consequences of a tourism system, which is the performance of that system. The consequences of the tourism network are the tourism destination performance as networks and alliances in tourism have been linked to improving tourist industry performance (March and Wilkinson, 2009).

A number of studies have been published in the tourism literature about application of network analysis to tourism management research. Ness, Aarstad, Haugland and Grønseth (2014) propose a role for inter-destination bridge ties in destination development. Pforr, Pechlaner, Volgger and Thompson (2014) show the effects of changes to a destination network and the importance of actors in the network to create a path to transform the destination and to meet the challenges of tourism. Kimbu and Ngoasong (2013) explored actors involved in tourism development in the Cameroon using social network analysis. Their analysis highlighted the challenges of coordinating the roles of public and private sector actors in tourism development and the empowerment of private sector and local communities in tourism policy, planning and development (Kimbu and Ngoasong, 2013). It would be important that advances are made in tourism research by utilizing a network perspective as this will build an understanding of tourism structures and patterns that improve the consequences of a tourism system.

Network analysis has been utilized to explore aspects of destination management. Hristov (2015) makes reference to Del Chiappa and Presenza's (2013) suggestion that little or no investigation has been conducted about the complete Destination Management Organization (DMO) network, and proposes a methodological framework that utilizes SNA to study DMOs. Hristov's (2015) framework involves three phases. The first phrase is qualitative and seeks to understand the forces that influence the emergence of the network. The second phase is quantitative and involves the capture of relational data about actors, roles and resource flows. The third phase is qualitative and seeks through interviews with DMO actors to resolve issues affecting network performance and informs interventions and implications for theory and practice in DMO management. Similarly, Jørgensen (2016) proposes combining quantitative and qualitative methodologies in a synergistic approach utilizing SNA. Thereby network measurements can form the basis for qualitative data exploration to reveal for instance reasons for a DMO receiving a high closeness centrality value, which is a measure of the degree a node is close to all other nodes in the network. Luthe and Wyss (2014) note the importance of backing quantitative results with qualitative data in a resilience context. They suggest that an integrated analysis will clarify the qualitative data relating to the functionality of ties and the quantitative data relating to the structure of ties.

Managing a destination network still requires attention (see Timur and Getz, 2008; Meriläinen and Lemmetyinen, 2011). Meriläinen and Lemmetyinen (2011, p. 26) define a destination network as:

> ... an inter-organizational, goal-oriented network embedded in the destination, comprising value-creating activities that are linked to each other through tourism business relationships, and that require resources of tourism companies or other organizations.

Thus, network analysis applied to tourism management research should be viewed as two levels that involve the mapping of a tourism system as the interaction of the tourist with the system when the tourist receives services, and the interaction of the host with the system when a management activity of service delivery occurs. On one hand, the tourist moves from one supplier to another within the tourism system as a tourism experience is created. On the other hand, the host moves through the tourism system to engage with activities to deliver services, such as an employee or manager. A host, who is a local, can also govern a tourist experience, such as destination policy, planning and implementation. The two levels work hand-in-hand as a destination network. On the whole, if both levels are mapped and combined then the human and non-human interactions within a tourism system can be understood as a pattern is determined. This work can be conducted for a small island as a test case and replicated to understand the different structures and patterns that result in certain consequences for a tourism system.

12.6 Conclusion

A dearth of papers exists about network analysis in island tourism research. While research about island environments in relation to sustainability has focused on the physical environment there has been neglect about the business of tourism and the benefits to be derived. Islands can be remote entities, such as in the Indian Ocean, and the nature of tourism industries warrants methodologies that take into account the entire structure of the tourism system to breakdown and rebuild or restructure the tourism industry for optimal results. Herein is the value of network analysis applied to island tourism research. Based on the challenges of island tourism research and the opportunity that island environments provide to improve the research process and understanding of the tourism phenomenon, a methodological framework that conducts network analysis can add value to the field.

Islands that have dominant tourism sectors are largely tourism systems. Nonetheless, the number of nodes and ties would be less than other elaborate tourism sectors spread across large countries and regions. Hence the research about tourism networks are ideal for island tourism research. The research will first start with a definition of the tourism system that would account for both external entities such as airlines and tour operators, and internal entities such as accommodation and attractions. In constructing the island tourism network, the direct ties are captured first and this will lead to the indirect ties. The next step would be to identify the interconnections by determining the connection between each node and this can be done through either secondary or primary research or both. The mapping of a tourism system for an island destination will involve both the business activities and also the non-business activities. The non-business activities surround the governance or management and steering of the tourism system so that it becomes high performing and provides benefits for its stakeholders. Such optimization can be analysed by characterizing the overall topology of the network structure (see Borgatti *et al.*, 2013) since for instance, if the network of the island tourism system is centralized (star-shaped) or decentralized (circular-shaped) has implications for the performance of the tourism system in terms of coordination and resource sharing. An example of this is the two network shapes in Chapter 6 of this volume by McLeod, Chambers and Airey (2017) wherein the circular shape of the Jamaica tourism policy network shows a separation between market and hierarchical actors and the more Y-shaped (from left to right) Trinidad and Tobago tourism policy network shows the limitation of those actors at the one end to influence those actors at the opposite end. Such results can be utilized by the Destination Management Organization to reformulate the network structure for optimal results.

Acknowledgements

The author acknowledges the support of the Faculty of Social Sciences, The University of the West Indies, Mona, Jamaica.

References

Baggio, R. (2010) Collaboration and coopertion in a tourism destination: a network science approach. *Current Issues in Tourism* 14(2), 183–189.

Baggio, R. and Sainaghi, R. (2016) Mapping time series into networks as a tool to assess the complex dynamics of tourism systems. *Tourism Management* 54(2016), 23–33.

Baggio, R., Scott, N. and Cooper, C. (2010a) Improving tourism destination governance: a complexity science approach. *Tourism Review* 65(4), 51–60.

Baggio, R., Scott, N. and Cooper, C. (2010b) Network science: a review focused on tourism. *Annals of Tourism Research* 37(3), 802–827.

Baldacchino, G. (2006) Warm versus cold water island tourism. *Island Studies Journal* 1(2), 183–200.

Belle, N. and Bramwell, B. (2005) Climate change and small island tourism: Policy maker and industry perspectives in Barbados. *Journal of Travel Research* 44(1), 32–41.

Benckendorff, P. and Zehrer, A. (2013) A network analysis of tourism research. *Annals of Tourism Research* 43, 121–149.

Bonetti, E., Petrillo, C.S. and Simoni, M. (2006) Tourism System Dynamics: A multi-level destination approach. In: Lazzeretti, L. and Petrillo, C.S. (eds) *Tourism Local Systems and Networking*. Elsevier, Oxford, UK, pp. 111–133.

Borgatti, S.P. and Halgin, D.S. (2011) On network theory. *Organization Science* 22(5), 1168–1181.

Borgatti, S.P., Everett, M.G. and Johnson, J.C. (2013) *Analyzing Social Networks*. SAGE Publications Ltd, London.

Bramwell, B., Higham, J., Lane, B. and Miller, G. (2017) Twenty-five years of sustainable tourism and the Journal of Sustainable Tourism: Looking back and moving forward. *Journal of Sustainable Tourism* 25(1), 1–9.

Brown, K., Cave, J., Moyle, B., Croy, G. and Weiler, B. (2010) Tourism interaction on islands: The community and visitor social exchange. *International Journal of Culture, Tourism and Hospitality Research* 4(2), 96–107.

Carlsen, J. (1999) A system approach to island destination management. *Systems Research and Behavioral Science* 16(4), 321–327.

Charlie, C. (2014) Collaborative environmental governance networks in small Indonesian island tourism destinations. PhD Thesis. Victoria University, Australia.

Coccossis, H. (2002) Island tourism development and carrying capacity. In: Apostolopoulos, Y. and Gayle, D.J. (eds) *Island Tourism and Sustainable Development, Caribbean, Pacific and Mediterranean Experiences*. Greenwood Publishing Group, Inc, Westport, Connecticut, pp. 131–144.

Dann, G. (1997) Paradigms in tourism research. *Annals of Tourism Research* 24(2), 472–474.

Davies, B. (2003) The role of quantitative and qualitative research in industrial studies of tourism. *International Journal of Tourism Research* 5(2), 97–111.

Echtner, C.M. and Jamal, T.B. (1997) The disciplinary dilemma of tourism studies. *Annals of Tourism Research* 24(4), 868–883.

Giddens, A. (1984) *The constitution of society: outline of the theory of structuration*. Polity, Cambridge, UK.

Granovetter, M.S. (1973) The strength of weak ties. *American Journal of Sociology* 78(6), 1360–1380.

Guba, E.G. (ed.) (1990) *The Paradigm Dialog*. Sage Publications Inc, Newbury Park, California.

Harrison, D. (2001) Tourism in small islands and microstates. *Tourism Recreation Research* 26(3), 3–8.

Hristov, D. (2015) Investigating DMOs through the lens of social network analysis: theoretical gaps, methodological challenges and practitioner perspectives. *Advances in Hospitality and Tourism Research* 3(1), 18–39.

Hummel, J. and van der Duim, R. (2016) SNV's modes of ordering: Organizing tourism as development practice. *Tourism Management* 57, 312–322.

Iaquinto, B.L. (2016) Strengths and weaknesses of using mixed methods to detect the sustainable practices of backpackers: a reflexive account. *Journal of Cleaner Production* 111, 479–486.

Jørgensen, M.T. (2016) Synergistic social network analysis: a synergistic approach to qualitative and quantitative network analysis. *Tourism Analysis* 21(6), 559–576.

Kimbu, A.N. and Ngoasong, M.Z. (2013) Centralised decentralisation of tourism development: a network perspective. *Annals of Tourism Research* 40, 235–259.

Ko, T.G. (2005) Development of a tourism sustainability assessment procedure: a conceptual approach. *Tourism Management* 26(3), 431–445.

Latour, B. (1996) On actor-network theory: a few clarifications. Available at: http://www.bruno-latour.fr/sites/default/files/P-67%20ACTOR-NETWORK.pdf (accessed 9 October 2017).

Latour, B. (2005) *Reassembling the Social: An introduction to Actor-Network Theory*. Claredon, Oxford, UK.

Law, J. (2007) *Actor network theory and material semiotics*. Available at: http://heterogeneities.net/publications/Law2007ANTandMaterialSemiotics.pdf (accessed 9 October 2017).

Leiper, N. (1979) The framework of tourism: Towards a definition of tourism, tourist, and the tourist industry. *Annals of Tourism Research* 6(4), 390–407.

Lim, C.C. and Cooper, C. (2009) Beyond sustainability: optimising island tourism development. *International Journal of Tourism Research* 11(1), 89–103.

Luthe, T. and Wyss, R. (2014) Assessing and planning resilience in tourism. *Tourism Management* 44, 161–163.

Manderscheid, K. (2016) Quantifying mobilities? Reflections on a neglected method in mobilities research. *Applied Mobilities* 1(1), 43–55.

March, R. and Wilkinson, I. (2009) Conceptual tools for evaluating tourism partnerships. *Tourism Management* 30(3), 455–462.

McLeod, M.T. (2005) Politics and Tourism in Tobago. MSc thesis, University of Surrey, UK.

McLeod, M. and McNaughton, M. (2016) Mapping an emerging open data ecosystem. *The Journal of Community Informatics* 12(2) (Special issue on Open Data for Social Change and Sustainable Development), 26–46.

McLeod, M., Vaughan, D.R. and Edwards, J. (2010) Knowledge networks in the tourism sector of the Bournemouth, Poole and Christchurch conurbation: preliminary analysis. *The Service Industries Journal* 30(8), 1651–1667.

McLeod, M., Chambers, D. and Airey, D. (2017) A comparative analysis of tourism policy networks. In McLeod, M. and Croes, R. (eds) *Tourism Management in Warm-water Island Destinations*. CAB International, Wallingford, UK.

McLeod, M., Lewis, E. and Spencer, A. (2017) Re-inventing, revolutionizing and transforming Caribbean tourism: multi-country tourism institutions and a research agenda. *Journal of Destination Marketing & Management* 6(2017), 1–4.

Meriläinen, K. and Lemmetyinen, A. (2011) Destination network management: a conceptual analysis. *Tourism Review* 66(3), 25–31.

Merinero-Rodríguez, R. and Pulido-Fernández, J.I. (2016) Analysing relationships in tourism: A review. *Tourism Management* 54, 122–135.

Miller, W.L. and Crabtree, B.F. (1994) Clinical research. In: Denzin, N.K. and Lincoln, Y. (eds) *Handbook of Qualitative Research*. Sage, Thousand Oaks, California, pp. 340–352.

Mitchell, R.E. and Reid, D.G. (2001) Community integration: Island tourism in Peru. *Annals of Tourism Research* 28(1), 113–139.

Ness, H., Aarstad, J., Haugland, S.A. and Grønseth, B.O. (2014) Destination development: the role of inter-destination bridge ties. *Journal of Travel Research* 53(2), 183–195.

Nunkoo, R. and Ramkissoon, H. (2010) Small island urban tourism: a residents' perspective. *Current Issues in Tourism* 13(1), 37–60.

Padget, E., Dimanche, F. and Mounet, J.-P. (2010) A tourism innovation case, an actor-network approach. *Annals of Tourism Research* 37(3), 828–847.

Pavlovich, K. (2003) The evolution and transformation of a tourism destination network: the Waitomo Caves, New Zealand. *Tourism Management* 24(2), 203–216.

Pearce, D.G. (2008) Tourism planning in small tropical islands: methodological considerations and development issues in Samoa. *Etudes caribbeenes*, 9–10. Available at http://etudescaribeennes.revues.org/1393 (accessed 9 October 2017).

Pechlaner, H., Volgger, M. and Herntrei, M. (2012) Destination management organizations as interface between destination governance and corporate governance. *Anatolia* 23(2), 151–168.

Pforr, C., Pechlaner, H., Volgger, M. and Thompson, G. (2014) Overcoming the limits to change and adapting to future challenges governing the transformation of destination networks in Western Australia. *Journal of Travel Research* 53(6), 760–777.

Ren, C., Jóhannesson, G.T. and Van der Duim, R. (2012) How ANT works. In: Van der Duim, R., Ren, C. and Jóhannesson, G.T. (eds) *Actor-Network Theory and Tourism: Ordering, Materiality and Multiplicity*. Routledge, Oxford, UK.

Scheyvens, R. and Momsen, J. (2008) Tourism in small island states: From vulnerability to strengths. *Journal of Sustainable Tourism* 16(5), 491–510.

Scott, N., Baggio, R. and Cooper, C. (2008) *Network Analysis and Tourism: From Theory to Practice*. Channel View Publications, Bristol, UK.

Scott, N., Cooper, C. and Baggio, R. (2008) Destination networks: Four Australian cases. *Annals of Tourism Research* 35(1), 169–188.

Senna, M.L.G.S., Dutra, V.C. and Aquino, A.R. (2015) Methodologies for management of sustainable tourism: a case study in Jalapão/Tocantins/Brazil. *International Journal of Social, Behavioral, Educational, Economic and Management Engineering* 9(8), 2646–2650.

Sharpley, R. (2012) Island tourism or tourism on islands? *Tourism Recreation Research* 37(2), 167–172.

Sharpley, R. and Ussi, M. (2014) Tourism governance in small island developing states (SIDS): the case of Zanzibar. *International Journal of Tourism Research* 16(1), 87–96.

Sheldon, P. (2005) The challenges to sustainability in island tourism. *Travel Industry Management Occasional Paper* 2005-01, October 2005. University of Hawaii, Hawaii.

Shih, H.-Y. (2006) Network characteristics of drive tourism destinations: An application of network analysis in tourism. *Tourism Management* 27(5), 1029–1039.

Slevitch, L. (2011) Qualitative and quantitative methodologies compared: Ontological and epistemological perspectives. *Journal of Quality Assurance in Hospitality and Tourism* 12(1), 73–81.

Timur, S. and Getz, D. (2008) A network perspective on managing stakeholders for sustainable urban tourism. *International Journal of Contemporary Hospitality Management* 20(4), 445–461.

Tribe, J. (1997) The indiscipline of tourism. *Annals of Tourism Research* 24(3), 638–657.

Tribe, J., Dann, G. and Tazim, J. (2015) Paradigms in tourism research: a trialogue. *Tourism Recreation Research* 40(1), 28–47.

Yang, J., Ge, Y., Ge, Q., Xi, J. and Li, X. (2016) Determinants of island tourism development: The example of Dachangshan Island. *Tourism Management* 55, 261–271.

Zappino, V. (2005) Caribbean tourism and development: An overview. Discussion Paper No. 65. European Centre for Development Policy Management, Italy.

Zhang, G.-H., Liu, J. and Wan, R. (2008) Functional zoning of island tourism in Qingdao. *Resources Science* 30(8), 1155–1161.

Index

Page numbers in **bold** refer to illustrations and tables